Tropical Forest Ecosystem Services in Improving Livelihoods For Local Communities

Zaiton Samdin • Norfaryanti Kamaruddin •
Sheriza Mohd Razali
Editors

Tropical Forest Ecosystem Services in Improving Livelihoods For Local Communities

Springer

Editors
Zaiton Samdin
School of Business and Economics
Universiti Putra Malaysia (UPM)
Selangor, Malaysia

Norfaryanti Kamaruddin
Institute of Tropical Forestry and Forest Products (INTROP)
Universiti Putra Malaysia (UPM)
Selangor, Malaysia

Sheriza Mohd Razali
Institute of Tropical Forestry and Forest Products (INTROP)
Universiti Putra Malaysia (UPM)
Selangor, Malaysia

ISBN 978-981-19-3341-7 ISBN 978-981-19-3342-4 (eBook)
https://doi.org/10.1007/978-981-19-3342-4

This Springer imprint is published by the registered company Springer Nature Singapore Pte Ltd.
The registered company address is: 152 Beach Road, #21-01/04 Gateway East, Singapore 189721, Singapore

Contents

About the Editors

Zaiton Samdin is an Associate Professor in the School of Business and Economics, Universiti Putra Malaysia, and currently a Deputy Director, Institute of Tropical Forestry and Forest Products (INTROP), Universiti Putra Malaysia. She received her PhD degree from the University of Exeter, United Kingdom. Her research interests cover a range of area in ecotourism, sustainable tourism, economic valuation and tourism economics which includes willingness to pay as well as valuation studies in national park and protected areas.

Norfaryanti Kamaruddin is a Senior Researcher at the Laboratory of Sustainable Bioresource Management, Institute of Tropical Forestry and Forest Products (INTROP) in Universiti Putra Malaysia. She received her BSc in Bioindustry majoring in Agriculture Economics and MSc degree in the Agribusiness. She earned her PhD degree on the subject of industrial strategy of agribusiness corporations. She is a team member of CoE of Biomass Valorisation, CIRAD-INTROP. Her research interests cover a range of area in corporate network analysis, bioresource management, and techno-economic assessment of wood-based industry. She is heavily involved in CIRAD-UPM Cooperation and an active team member of Sustainable Agricultural Landscapes in Southeast Asia (SALSA) platform.

Sheriza Mohd Razali was born in Malaysia on February 16, 1978. She received her BSc in Forest Management and MSc degree in Application of Remote Sensing and GIS from Universiti Putra Malaysia, and PhD in Biodiversity and Environmental Management (Remote Sensing and GIS) from Universidad De Murcia, Murcia, Spain. She is currently a Research Officer in the Institute of Tropical Forestry and Forest Products, Universiti Putra Malaysia. She was Research Officer in Malaysian Remote Sensing Agency from 2002 to 2003. She is a member of the Fundamental Research Grant Scheme (FGRS), Transdisciplinary Research Grant Scheme (TGRS), and NAPC: Network ASEAN Peat Swamp Forest Communities projects from 2018 until now. She is currently the project leader for Consultancy for Hydrological Modelling Study for the Ulu Muda FR, collaboration between WWF Malaysia, UPM and Universiti Malaysia Terengganu. Her research interests include remote sensing and GIS. She is an expert in ArcGIS, ERDAS and ENVI software. She has published 18 journal papers, a book, 5 book chapters, 18 proceedings and 7 other publications.

1 Conservation of Tropical Forest for the Well-Being of Community

Wan Norhidayah W. Mohamad

Abstract

Tropical forests host a large population of biodiversity that play a crucial role in global climate regulation. Besides that, it represents a foundation for the provision of ecosystem services such as clean air and water, valuable timber and animal and plant resources with high commercial and cultural value. However, tropical forests are facing great pressure as a result of increasing human exploitation. If the world's tropical forests are destroyed, then many of the biodiversity species will be lost along with them. Not only that, but the local community also loses the natural system that performs valuable services which is important for the continuity of human's life. The balance of economic growth and conservation of biodiversity and its components including tropical forest must be achieved. Having said this, the ongoing action in conserving our valuable resources of tropical forest is important especially to support the well-being of the local community. Overall, this chapter discusses the importance of tropical forests, threats, conservation action as well as the economic value and economic valuation techniques that can be used to put an economic value on these natural resources.

Keywords

Tropical forest · Conservation · Biodiversity · Economic valuation

W. N. W. Mohamad (✉)
School of Business and Economics, Universiti Putra Malaysia (UPM), Serdang, Selangor Darul Ehsan, Malaysia
e-mail: w_norhidayah@upm.edu.my

Z. Samdin et al. (eds.), *Tropical Forest Ecosystem Services in Improving Livelihoods For Local Communities*, https://doi.org/10.1007/978-981-19-3342-4_1

1.1 Introduction

The world's tropical rainforests are divided into five separate biogeographical and ecological regions: Madagascar, Africa, Asia, the Neotropics and New Guinea whereby each region has its unique biota and interactions (Corlett & Primack, 2006). Brazil, Peru, the Democratic Republic of Congo (DRC), Indonesia and Colombia are the countries having the most tropical forest (Butler, 2020). The Central African Republic, the Republic of Congo, Ecuador, Bolivia, Cameroon, India, Gabon, Papua New Guinea, Guyana, Mexico, Laos, Malaysia, Myanmar and Venezuela are among the countries with extensive expanses of the rainforest. In total, the overall forest area is 4.06 billion hectares, and forests currently cover 30.8% of the global land area (FAO & UNEP, 2020). Figure 1.1 shows the worldwide distribution of forests in million hectares and the percentage of the world's forests.

Tropical forests are one of the greatest diverse, besides the most dynamic, biomes on the planet, where tremendous amounts of carbon are accumulated and stored (Baccini et al., 2017; Sullivan et al., 2017; Di Marco et al., 2018). For the role they serve in supporting life on this planet, tropical rainforests are a priceless asset to all living organisms. These luxuriant forests are conquered by broad-leaved plants that yield a dense upper canopy and include a rich variety of flora and fauna in damp tropical highlands and lowlands surrounding the Equator. Tropical rainforests do not

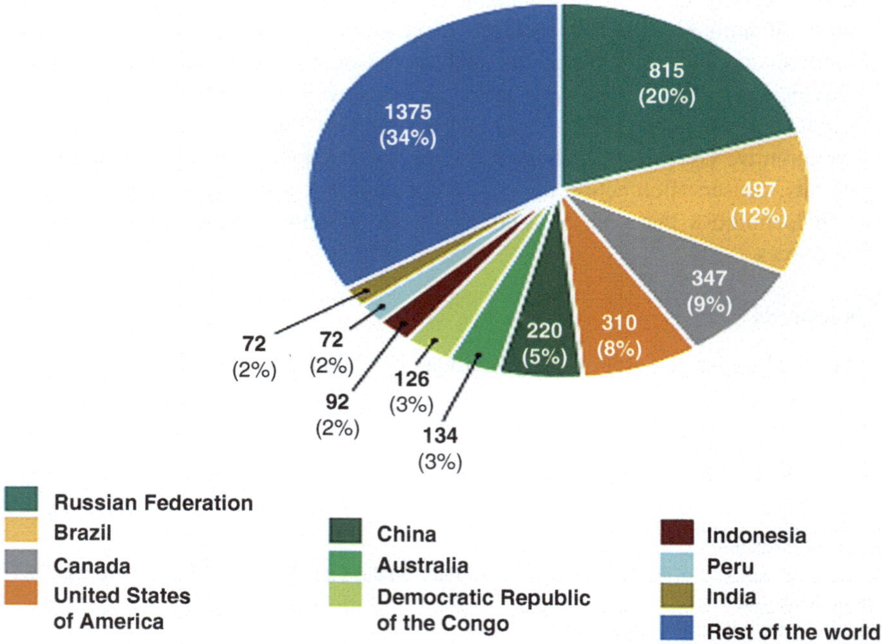

Fig. 1.1 Worldwide distribution of forests. (Source: FAO and UNEP (2020))

all flourish in areas with abundant and consistent rainfall. For example, the weather in the dry rainforests of north-eastern Australia is broken by a dry season, which decreases annual precipitation.

Tropical rainforest, sometimes known as the primary tropical rainforest, is constantly changing and evolving. More than one-third of the world's forests, or 34%, are primary forests, defined as naturally regenerated forests of native tree species with no visible signs of human activity and ecological processes that are not significantly disturbed (FAO & UNEP, 2020). Nowadays, disturbance arrays in tropical environments are varying due to factors, for example, increased land use pressure for economic development (e.g. agriculture, mining, logging, ranching) and natural disaster such as altered patterns in hurricanes. The process of seeking, cutting, arranging and removing rainforest wood, even in so-called selective logging, can be onerous (Nepstad et al., 1999). According to the FAO and UNEP (2020), since 1990, the areas of forest that have been destroyed due to transformation to other land uses are 420 million ha.

Tropical rainforests are an important forest habitat that has trees that can be hundreds of years old or even thousands of years old. They retain more carbon than other types of forests and are essential for biodiversity conservation. Tropical rainforests are home to a variety of creatures, including orangutans, mountain gorillas, jaguars and tigers. Once these woods have been cut down, it is unlikely that they will ever return to their original state. This chapter aims to discuss the importance of tropical forests and conservation action to protect the forest and also provide a briefing introduction about the economic valuation techniques that can be used to put an economic value on these natural resources.

1.2 The Importance of Tropical Forest

The importance and benefit tropical forests provide to human well-being can be classified into three: environmental, economic and social benefits (see Fig. 1.2). In terms of economic benefits, forests provide humans with wood products (e.g. timber, lumber, fuelwood) and non-wood products (e.g. food). Social benefits gained from forest resources are recreation, aesthetic and spiritual values. Forests also provide a broad range of environmental benefits such as carbon dioxide absorbers, providing clean air and water, etc.

1.2.1 Forests Are a Source of Food, Raw Materials and Medicine

Tropical forests serve as both a home for all living things and a source of nutrition and lumber for hundreds of millions of people, as well as a barrier against extreme poverty (Edwards et al., 2019). Numerous tropical rainforest plants have therapeutic properties, particularly stems, roots, shoots, leaves and tubers. More than 28,000 plant species have been identified as having therapeutic properties, with many of them being found in forest habitats (FAO & UNEP, 2020). The forest also has a lot

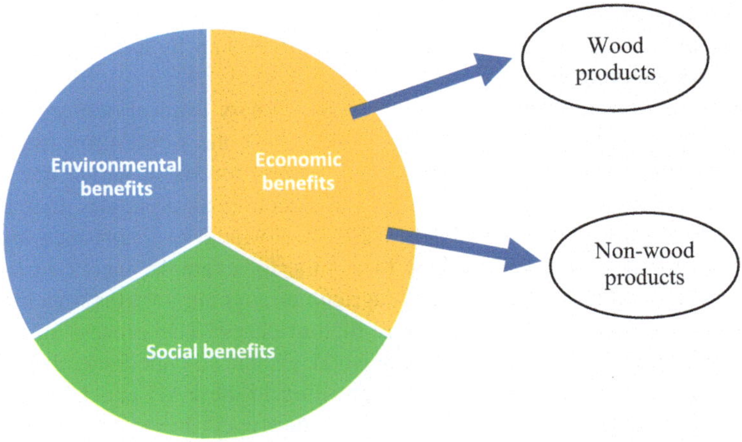

Fig. 1.2 The benefits of forest

of high-quality hardwoods such as merbau, meranti, cengal and keruing. All of these timber resources are worth a lot of money to the government. Forests also provide non-timber forest product source with a variety of applications, for example, rattan, mengkuang, bamboo and many more others.

1.2.2 Forest as Natural Ground Cover

Forests' natural features provide numerous environmental benefits. Forests serve as both a natural ground cover and a barrier against water and wind erosion. Grass cover, for example, is useful for controlling pollution and soil erosion (Zhang et al., 2020). By grasping the earth with its roots, the tree trunk acts as a binder. The forest also acts as a sponge, soaking precipitation and slowly releasing it into the streams, preventing floods. It will also serve as an indirect filter, assessing the cleanliness and clarity of the water. The quality and quantity of rivers are affected by the hydrological cycle or circulation that happens in forest environments (Koralay & Kara, 2018). Forest ecosystems operate as water regulators and restrict the flow rate of water in terms of quantity. At the same time, the tree canopy is capable of preventing or absorbing rainwater. This could all help to protect the environment from pollution and avoid disasters like flash floods and landslides.

1.2.3 Forests as Earth Weather Stabilisers

Tropical rainforests provide over 40% of the world's oxygen supply (World Wildlife Fund for Nature (WWF), 2013). They take in carbon dioxide and release oxygen at the same time that plants do photosynthesis. Animal and plant respiration, as well as other activities that result in oxygen intake and carbon dioxide generation through the carbonisation process, balance this out. Long-term deforestation will limit the conversion of carbon dioxide to organic carbon in the atmosphere. As a result, higher carbon dioxide levels in the atmosphere will induce global warming. It will have an impact on the reduction of gas levels in the atmosphere, as well as the thickness of the ozone layer.

1.3 Deforestation of Tropical Forest

Deforestation has undeniably detrimental consequences on the environment. Forecasts suggest that the combination of deforestation and climate change might cause a loss of up to 58% of Amazonian tree species richness, with 53% of these species at risk of extinction, even if protected areas may assist to mitigate these effects (Gomes et al., 2019). This highlights the necessity for a better conservation strategy in dealing with the joint impacts of land use and climate change (Struebig et al., 2015).

Despite the growing international concern, rainforests continue to be destroyed at an alarming rate, owing primarily to increased agricultural demand (Aide et al., 2013; Capellesso et al., 2021). Due to a recent pattern of PA downsizing, downgrading and degazettement, the future of tropical forest protected areas is dubious (Kroner et al., 2019), combined with ongoing trends of degradation and severe human force within PAs (Jones et al., 2018). The rate of forest loss varies widely between countries, although deforestation has been widespread in Asia, Africa and Latin America (World Resource Institute (WRI), 2021).

Figure 1.3 shows the location of tropical forests and deforested areas in the early twenty-first century around the world. It can be seen from the figure that deforestation has taken place in all forest areas. If deforestation continues to happen, this will reduce the amount of forest area available and becomes a threat to all life in the universe. Without a doubt, tropical forest destruction has global consequences, including biodiversity loss (Strassburg et al., 2010; Symes et al., 2018), along with the release of the whole carbon into the atmosphere (Barni et al., 2015; Brinck et al., 2017).

Figure 1.4 shows the tropical forest lost from 2002 to 2020. It can be seen from the figure that 2016 recorded the highest tropical loss with approximately 6.2 million ha, followed by the loss of 5 million ha in 2017 and approximately 4.2 million ha loss in 2020. The loss of 4.2 million ha, almost the Netherlands size, took place in humid tropical primary forests, which are particularly vital for biodiversity and carbon storage. According to the World Resource Institute (WRI) (2021), the main causes of this loss are increased deforestation and fires in the Amazon, which is the world's largest rainforest. The annual carbon emissions from this main forest loss are comparable to 570 million cars, which is more than twice the number of cars on the road in the United States.

Figure 1.5 presents the top ten countries for primary forest loss in 2020. Brazil reported the highest loss, with a total of 1.7 million ha loss in 2020, and this number increased by 25% in 2020 compared to 2019. According to the WRI report, fires and

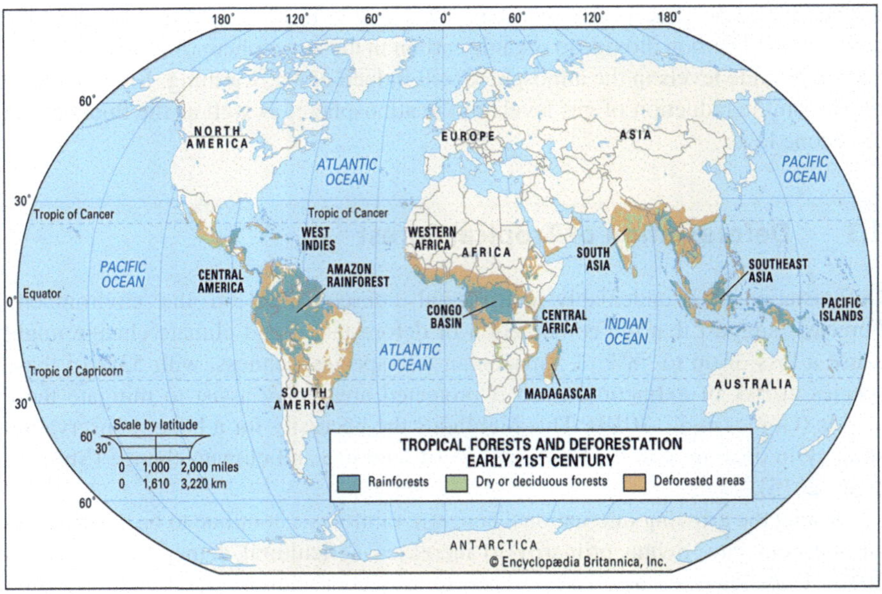

Fig. 1.3 Tropical forests and deforestation early twenty-first century. (Source: Retrieved from https://www.britannica.com/science/tropical-rainforest)

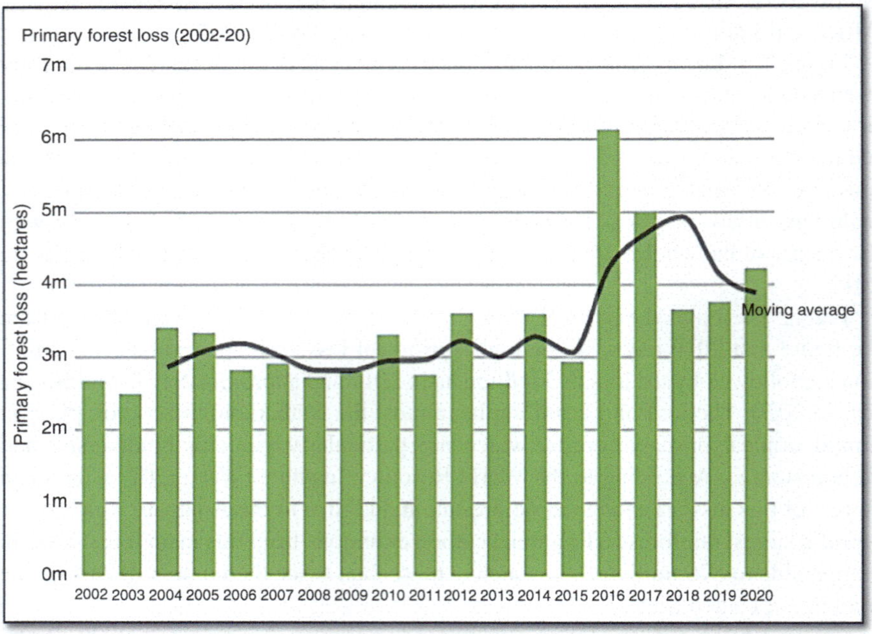

Fig. 1.4 Tropical forests lost from 2002 to 2020. (Source: World Resource Institute (WRI) (2021))

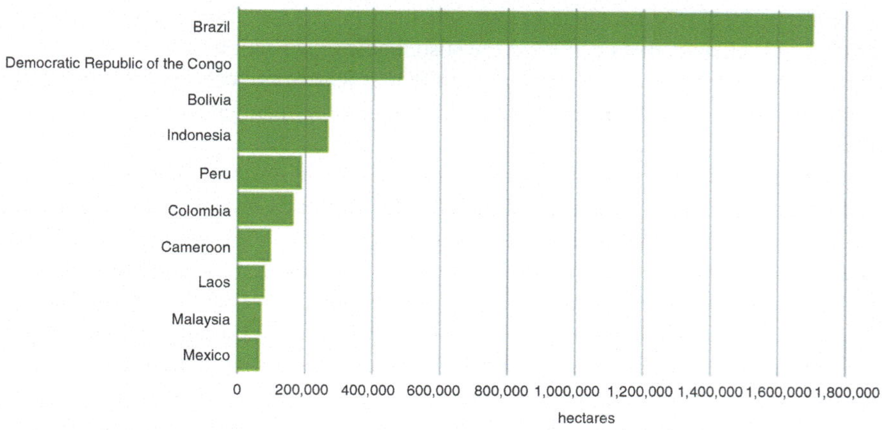

Fig. 1.5 Top 10 countries for 2020 primary forest loss. (Source: World Resource Institute (WRI) (2021))

clear-cutting are the main causes of forest loss in Brazil. The other reason for this loss is the small-scale shifting agriculture and wood energy demands that happen in the Democratic Republic of the Congo and southern part of Colombia. Meanwhile, in Australia, extreme weather was the cause of primary forest loss with climate change likely to make fire-prone conditions more common in the future.

Tropical rainforest deforestation has a negative global impact due to the loss of essential ecosystem services and renewable resources, as well as species extinction and the loss of carbon sinks. However, under certain circumstances, this degradation can be slowed, stopped or even reversed. Most people agree that something needs to be done about the problem, but the solution is not as simple as erecting walls around the surviving rainforests or outlawing the timber trade. If forests are fully cut off from use and development, economic, political and social forces will make it impossible for them to survive. Therefore, efforts to restore and conserve existing forests through specific conservation actions are very important to be highlighted at all times.

1.4 The Importance of Tropical Forest Conservation

These days tropical rainforests are vanishing from the face of the globe. Various forms of forest conservation measures are being introduced and implemented to preserve our tropical rainforest. Besides traditional mechanisms like protected areas (PAs), other initiatives are also being carried out including payments for environmental services (PES), certification schemes and development programmes.

The core concept of PES is that landowners and managers are compensated by beneficiaries of these services for the provision of specific ecosystem services or for a specific forest management plan that generates the wanted ecosystem services

(Prokofieva, 2016). Meanwhile, eco-certification or eco-labelling is an example of certification schemes that encourage the sustainable exploitation of forest resources. This scheme is founded on the assumption that at least some consumers will opt to buy things that are made in an ecologically responsible manner (Nunes & Riyanto, 2005).

Another example is the Tropical Forest Conservation Act (TFCA) that was implemented by the US government in 1998. In 2019, this most successful international and environmental policy act was updated to the Tropical Forest and Coral Reef Conservation Act. In a nutshell, this act gives qualified developing countries alternatives for resolving certain US government obligations while simultaneously generating cash in local currency to help fund tropical forest conservation activities.

Tropical forests support at least two-thirds of Earth's terrestrial biodiversity (Gardner et al., 2009), despite covering only 6–7% of the land surface (Dirzo & Raven, 2003). Since the mid-twentieth century, the tropical forest area designated for biodiversity conservation has grown (Watson et al., 2014). Regardless of these efforts, PAs in the tropics are experiencing biodiversity loss (Laurance et al., 2012; Tranquilli et al., 2014).

As a result, what should be done? The path out must be realistic, not idealistic and based on the establishment of a conservation technique based on the premise of long-term usage and development of tropical rainforests. Aside from tropical rainforest development, determinations to integrate and restore tainted forest areas, besides the formation of PAs, are critical to preserving tropical rainforests for the long-term advantages they can offer humanity.

Historic techniques of rainforest protection have been proven to be ineffective, as seen by the rapid rate of deforestation. Protected areas can aid in the conservation of tropical forests, although they are frequently underfunded and so under-protected (e.g. Inogwabini et al., 2005). Closing off woods as untouchable parks and reserves has neither enhanced the quality of life or economic prospects for rural impoverished nor has it discouraged illicit loggers and developers from cutting forests. The difficulty with the traditional park strategy to sustain wildlands in developing nations is that it does not provide sufficient economic incentives for forest preservation.

Tropical rainforests will only persist as efficient ecosystems if they can be demonstrated to deliver measurable economic advantages. Residents and the responsible bodies must understand financial returns to defend the expenditures of preserving parks and foregoing revenue from commercial activities within the boundaries of protected areas. If we want to save the world's surviving tropical rainforests, we cannot just stop the timber trade or create reserves. Destruction of forests can only be stopped by understanding and eliminating the underlying socioeconomic and political factors that contribute to it. After the difficulties have been identified, a decision about what should be accomplished can be made.

Putting the economic value on forest resources is one of the efforts to justify the importance of conservation of these natural resources. A tropical forest's worth is derived not only from the market prices for its direct uses but also from other non-tradable uses of these resources. An economic valuation study, which will be explored in the following subtopic, can reveal this worth.

1.5 The Economic Valuation of Tropical Forest

Environmental advantages of forests have been underrated by decision-makers because they are difficult to evaluate and many are outside of typical markets and pricing systems. Due to the advancements in economic valuation methodologies during the last decade, forest benefits may now be computed and expressed in monetary terms. Economic valuation is the measurement of people's preferences for an environmental good or against an environmental bad (Pearce, 1991). As a result, valuation refers to people's preferences. Resulting valuations are in monetary terms as they are based on the method of preference revelation, i.e. asking what people are ready to pay or inferring their WTP through other means.

Before we go further on the economic valuation techniques, it would be beneficial to value the goods and services offered by tropical forests. Figure 1.6 shows the total economic value (TEV) of tropical forests. Following Pearce (1991), TEV of tropical forest could be considered consisting of use value and non-use value. Specifically, use value can be further divided into four: direct use value, indirect use value, option

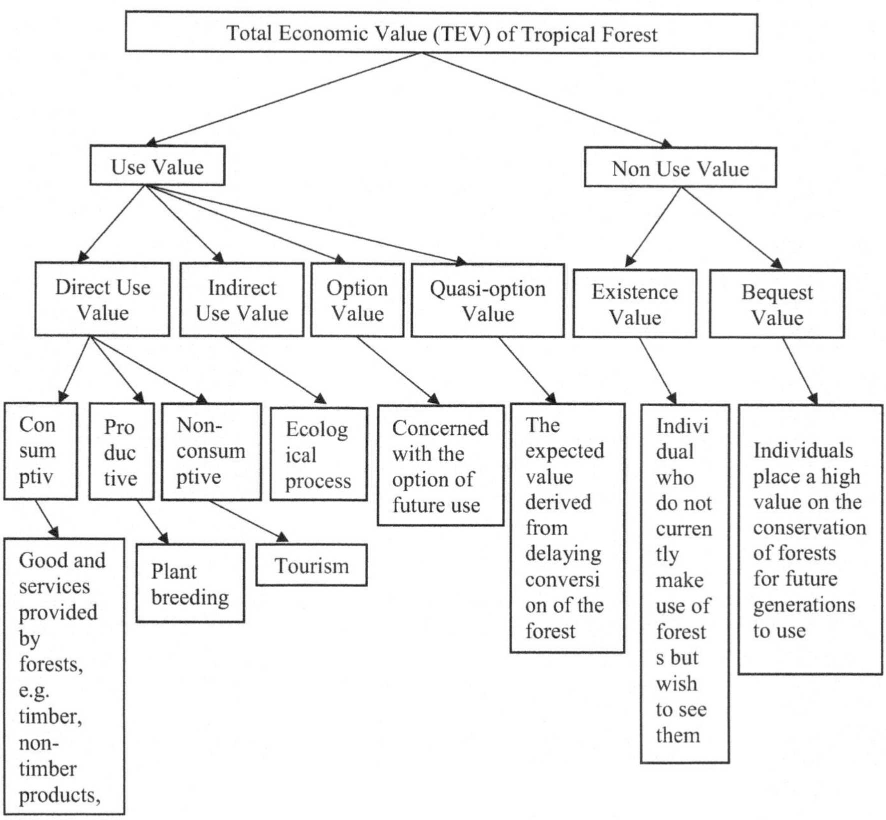

Fig. 1.6 Total economic value of tropical forest. (Source: Pearce (1991))

value and quasi-option value. Meanwhile, non-use value can be divided into two: existence value and bequest value.

In brief, direct value refers to goods and services provided by the forest, for example, timber and non-timber products, recreation, medicines, etc. Indirect use value refers to the ecological services and function of the forest in terms of facilitating nutrient cycling, carbon fixing, watershed protection, etc. Option value refers to the future use of both direct and indirect use values, for instance, the future value of the drug. Quasi-option values refer to the expected value derived from delaying the conversion of the forest today. Existence value which is a non-use value is concerned with viewing forests as objects of inherent value that need to be conserved. Bequest value refers to persons assigning a high value to the conservation of forest for future generations.

In the past, land use and investment decisions were made with little awareness of tropical forest economic value. Forests were once thought to be only economically valuable if they could support commercial timber or wood production. This was used to calculate the value of forest goods and services to household production, the profitability of a project, sectoral productivity and national economic indicators.

Unsurprisingly, economic policy tools and analyses of forest management alternatives revealed a strong preference for commercial extraction, agricultural clearance or alteration for other ostensibly profitable development options. Forest conservation and sustainable management appeared to have little economic benefits, whereas forest degradation and loss appeared to have few economic costs. Tropical forests, on the other hand, often give economic benefits in addition to commercial wood and timber products. They also offer sustenance commodities and environmental services, which are regularly worth significantly more money. Traditionally, economists have found it challenging to value or express these non-market advantages in monetary terms. As a result, they are frequently left out of decision-making.

Nonetheless, as economic valuation methods have advanced and people's requirements and expectations have changed, there has been a growing recognition of the significance of such values to business profits and trade, national economic welfare and household consumption and production. Simultaneously, it has become obvious that to effectively assess the whole range of social, economic and environmental trade-offs implied by alternative forest land use and management alternatives, it is necessary to be able to describe these broader forest values in quantifiable economic terms.

To evaluate the use and non-use values of tropical forests, a variety of economic valuation techniques, for instance, contingent valuation method (CVM), choice experiment (CE) and travel cost method (TCM) have been applied. All these methods are structurally different, and each of the methods has its advantages. Table 1.1 presents past studies that apply numerous economic valuation techniques to estimate the economic value of tropical forest conservation.

Table 1.1 Past studies on forest conservation using economic valuation techniques

Title and researchers	Purpose of study	Methods	Main finding
Timber or carbon? Evaluating forest conservation strategies through a discrete choice experiment Bocci et al. (2020)	To examine the trade-off between carbon storage and timber production in the Maya biosphere reserve Forest concessions, Guatemala	CE	Contracts that emphasise carbon storage are preferred by households, rather than timber harvesting. The households also want an entree to forests so they may gather non-timber forest products
Willingness to pay for forest conservation in Ecuador: results from a nationwide contingent valuation survey in a combined "referendum"—"Consequential open-ended" design Gordillo et al. (2019)	To evaluate preferences and willingness to pay of households for the suggested forest conservation programme in Ecuador, targeting at preventing deforestation	CVM	Approximately 98% of surveyed households reflect the suggested programme is worth supporting. The monthly mean WTP per household lies between $3.17 and $6.18 for forest conservation
Does the economic benefit of biodiversity enhancement exceed the cost of conservation in planted forests? Yao et al. (2019)	To evaluate the cost and benefits of conserving the New Zealand brown kiwi, an iconic yet endangered bird species that lived in planted forests	CE	The value of the proposed biodiversity conservation initiative at the national level can be more than 100 times higher than the overall cost of the programme
Valuing conservation benefits of disease control in wildlife: a choice experiment approach to bovine tuberculosis management in New Zealand's native forests Tait et al. (2017)	To value the native forest biodiversity benefits of bovine tuberculosis (TB)-related possum control	CE	People place substantial value on the most observable biodiversity benefits of TB possum control, for example, the presence of native birds and improved forest canopies
Traditional village forest landscapes: Tourists' attitudes and preferences for conservation Chen et al. (2017)	To examine a traditional village tree landscape that offers tourists an integrated experience of nature, history and local culture at the Ryukyu Islands	CVM	Respondents under 40 years old or travelling with their family were more likely to donate to the tree landscape conservation fund, whereas those with a greater household income or higher education level were less likely to donate to the village tree landscape conservation fund
Using contingent valuation to determine	To assess the conservation value of two	CVM	Australian visitors to Vanuatu would be willing

(continued)

Table 1.1 (continued)

Title and researchers	Purpose of study	Methods	Main finding
Australian tourists' values for forest conservation in Vanuatu Flatley and Bennett (1996)	rainforest regions in the Republic of Vanuatu		to pay roughly $403,000 per year to safeguard the rainforest locations, and that this amount would be enough to cover the expenses of lease agreements to ensure that logging activities were prohibited in these regions

1.6 Conclusion

In conclusion, the objective of this chapter is to highlight the importance of tropical forests, threats, conservation action as well as the economic value and economic valuation techniques that can be used to put an economic value on these natural resources. We rely on the forest for almost everything from the air we breathe to the food we eat. In addition to providing a habitat for animals and a source of income for humans, forests preserve watersheds, prevent soil erosion and mitigate climate change. We continue to let forests disappear despite our reliance on them.

Many social and economic forces such as mining and overexploitation for agricultural development purposes endangered tropical forests and their biodiversity. Forest conservation policies rely on the ability to demonstrate their economic benefit as well as social indicators that involve local community participation. Estimating the direct use value of tropical forest products, indirect values from environmental benefits and services and option and existence values has all been used in research to quantify the entire economic value of forests. Many indigenous tribes' social and economic well-being has been jeopardised by forest threats; their survival and forest conservation can frequently be achieved by maintaining traditional land tenure and environmental management systems.

The truth is that uncontrolled deforestation of tropical forests will have a long-term harmful impact on humanity since forests play such an important role in human life, particularly in terms of health and quality of life. The relationship between the environment and society is close; in fact, the two are mutually dependent. However, if the environment, particularly tropical forests, is destroyed, society as a whole will suffer losses. Therefore, high awareness of each individual in preserving and conserving forests is important to maintain human well-being.

References

Aide, T. M., Clark, M. L., Grau, H. R., López-Carr, D., Levy, M. A., Redo, D., Bonilla-Moheno, M., Riner, G., Andrade-Núñez, M. J., & Muñiz, M. (2013). Deforestation and reforestation of Latin America and the Caribbean (2001–2010). *Biotropica, 45*, 262–271.

Baccini, A., Wlaker, W., Carvalho, L., Farina, M., Sulla-Menashe, D., & Houghton, R. A. (2017). Tropical forests are a net carbon source based on aboveground measurements of gain and loss. *Science, 358*, 230–234.

Barni, P. E., Fearnside, P. M., & Graca, P. M. L. (2015). Simulating deforestation and carbon loss in Amazonia: Impacts in Brazil's Roraima state from reconstructing highway BR-319 (Manaus-Porto Velho). *Environmental Management, 55*, 259–278.

Bocci, C., Sohngen, B., Lupi, F., & Milian, B. (2020). Timber or carbon? Evaluating forest conservation strategies through a discrete choice experiment. *Ecological Economics, 171*, 106601.

Brinck, K., Fischer, R., Groeneveld, J., Lehmann, S., Paula, M. D., Pütz, S., Sexton, J. O., Song, D., & Huth, A. (2017). High resolution analysis of tropical forest fragmentation and its impact on the global carbon cycle. *Nature Communications, 8*, 14855.

Butler, R. A. (2020). *Rainforest information*. Retrieved from https://rainforests.mongabay.com

Capellesso, E. S., Cequinel, A., Marques, R., Sausen, T. L., Bayer, C., & Marques, M. C. M. (2021). Co-benefits in biodiversity conservation and carbon stock during forest regeneration in a preserved tropical landscape. *Forest Ecology and Management, 492*, 119222.

Chen, B., Nakama, Y., & Zhang, Y. (2017). Traditional village forest landscapes: Tourists' attitudes and preferences for conservation. *Tourism Management, 59*, 652–662.

Corlett, R. T., & Primack, R. B. (2006). Tropical rainforests: Why cross-continental comparisons are needed. *Trends in Ecology and Evolution, 21*, 104–110.

Di Marco, M., Watson, J. E., Currie, D. J., Possingham, H. P., & Venter, O. (2018). The extent and predictability of the biodiversity–carbon correlation. *Ecology Letters, 21*(3), 365–375.

Dirzo, R., & Raven, P. H. (2003). Global state of biodiversity and loss. *Annual Review Environmental Resource, 28*, 137–167.

Edwards, D. E., Socolar, J. B., Mills, S. C., Burivalova, Z., Koh, L. P., & Wilcove, D. S. (2019). Conservation of tropical forests in the anthropocene. *Current Biology, 29*, R1008–R1020.

FAO & UNEP. (2020). *The state of the world forest*. FAO & UNEP.

Flatley, G. W., & Bennett, J. W. (1996). Using contingent valuation to determine Australian tourist's values for forest conservation in Vanuatu. *Economic Analysis and Policy, 26*(2), 111–127.

Gardner, T. A., Barlow, J., Chazdon, R., Ewers, R. M., Harvey, C. A., Peres, C. A., & Sodhi, N. S. (2009). Prospects for tropical forest biodiversity in a human-modified world. *Ecological Letters, 12*, 561–582.

Gomes, V. H. F., Vieira, I. C. G., Salomão, R. P., & ter Steege, H. (2019). Amazonian tree species threatened by deforestation and climate change. *Nature Climate Change, 9*, 547–553.

Gordillo, F., Elsasser, P., & Günter, S. (2019). Willingness to pay for forest conservation in Ecuador: Results from a nationwide contingent valuation survey in a combined "referendum"—"Consequential open-ended" design. *Forest Policy and Economics, 105*, 28–39.

Inogwabini, B.-I., Ilambu, O., & Gbanzi, M. A. (2005). Protected areas of the Democratic Republic of Congo. *Conservation Biology, 19*, 15–22.

Jones, K. R., Venter, O., Fuller, R. A., Allan, J. R., Maxwell, S. L., Negret, P. J., & Watson, J. E. M. (2018). One-third of global protected land is under intense human pressure. *Science, 360*, 788–791.

Koralay, N., & Kara, O. (2018). Forestry activities and surface water quality in a watershed. *European Journal of Forest Engineering, 4*(2), 70–82.

Kroner, R. E. G., Qin, S., Cook, C. N., Krithivasan, R., Pack, S. M., Bonilla, O. D., Cort-Kansinally, K. A., Coutinho, B., Feng, M., Martínez Garcia, M. I., et al. (2019). The uncertain future of protected lands and waters. *Science, 364*, 881–886.

Laurance, W. F., Carolina Useche, D., Rendeiro, J., et al. (2012). Averting biodiversity collapse in tropical forest protected areas. *Nature, 489*, 290–294.

Nepstad, D., Verissimo, A., & Alencar, A. (1999). Large-scale impoverishment of Amazonian forests by logging and fire. *Nature, 398*, 505–508.

Nunes, P. A. L. D., & Riyanto, Y. E. (2005). Information as a regulatory instrument to price biodiversity benefits: Certification and ecolabeling policy practices. *Biodiversity and Conservation, 14*, 2009–2027.

Pearce, D. W. (1991). *Economic valuation and the natural world*. Earthscan Publications.

Prokofieva, I. (2016). Payments for ecosystem services—The case of forests. *Current Forestry Report, 2*, 130–142.

Strassburg, B. B. N., Kelly, A., Balmford, A., Davies, R. G., Gibbs, H. K., Lovett, A., Miles, L., Orme, C. D. C., Price, J., Turner, R. K., & Rodrigues, A. S. L. (2010). Global congruence of carbon storage and biodiversity in terrestrial ecosystems. *Conservation Letters, 3*, 98–105.

Struebig, M. J., Wilting, A., Gaveau, D. L. A., Meijaard, E., Smith, R. J., Consortium, T.B.M.D, Fischer, M., Metcalfe, K., & Kramer-Schadt, S. (2015). Targeted conservation to safeguard a biodiversity hotspot from climate and land-cover change. *Current Biology, 25*, 372–378.

Sullivan, M., Talbot, J., Lewis, S., et al. (2017). Diversity and carbon storage across the tropical forest biome. *Scientific Reports, 7*, 39102.

Symes, W. S., Edwards, D. P., Miettinen, J., Rheindt, F. E., & Carrasco, L. R. (2018). Combined impacts of deforestation and wildlife trade on tropical biodiversity are severely underestimated. *Nature Communications, 9*, 4052.

Tait, P., Saunders, C., Nugent, G., & Rutherford, P. (2017). Valuing conservation benefits of disease control in wildlife: A choice experiment approach to bovine tuberculosis management in New Zealand's native forests. *Journal of Environmental Management, 189*, 142–149.

Tranquilli, S., Abedi-Lartey, M., Abernethy, K., Amsini, F., Asamoah, A., Balangtaa, C., et al. (2014). Protected areas in tropical Africa: Assessing threats and conservation activities. *PLoS One, 9*(12), e114154.

Watson, J. E. M., Dudley, N., Segan, D. B., & Hockings, M. (2014). The performance and potential of protected areas. *Nature, 515*, 67–73.

World Resource Institute (WRI). (2021). *Global forest review*. Retrieved August, 2021, from https://research.wri.org/gfr/forest-pulse

World Wildlife Fund for Nature (WWF). (2013). Retrieved October, 2021, from https://wwf.panda.org/wwf_news/?207989/Life-sustaining-value-of-Earths-forests-highlighted-on-International-Day-of-Forests

Yao, R. T., Scarpa, R., Harrison, D. R., & Burns, R. J. (2019). Does the economic benefit of biodiversity enhancement exceed the cost of conservation in planted forests? *Ecosystem Services, 38*, 100954.

Zhang, N., Zhang, Q., Li, Y., Zeng, M., Li, W., Chang, C., Xu, Y., & Huang, C. (2020). Effect of groundcovers on reducing soil erosion and non-point source pollution in citrus orchards on red soil under frequent heavy rainfall. *Sustainability, 12*, 1146.

Forest Pathology in Ecosystem Services 2

Dzarifah Zulperi, Rohaya Ibrahim, Anysia Hedy Ujat, Adam Zafdri Md Zali, Yasmin Ja'afar, and Daljit Singh Karam

Abstract

Ecosystem services are processes in which the environment supplies benefits to humans and are classified into four major categories, including provisioning services, supporting services, regulating services, and cultural services. The increasing number of invasive pests and pathogens entering Malaysia signifies that the threat to forests is escalating along with climate change and globalization. In recent years, massive epidemics of forest diseases have devastated natural ecosystems and landscapes valued for timber and extended benefits to the community. Thus, an impeccable balance is required between food production to fulfill the global survival and maintenance of the other services supplied by the ecosystem services. This section reviews and discusses how the forest pathology system influences ecosystem services, specifically its effects on trees and forest diseases.

D. Zulperi (✉)
Laboratory of Sustainable Bioresource Management, Institute of Tropical Forestry and Forest Products (INTROP), Universiti Putra Malaysia (UPM), Serdang, Selangor, Malaysia

Department of Plant Protection, Faculty of Agriculture, Universiti Putra Malaysia (UPM), Serdang, Selangor, Malaysia
e-mail: dzarifah@upm.edu.my

R. Ibrahim · A. H. Ujat · A. Z. Md Zali · Y. Ja'afar
Laboratory of Sustainable Bioresource Management, Institute of Tropical Forestry and Forest Products (INTROP), Universiti Putra Malaysia (UPM), Serdang, Selangor, Malaysia

D. S. Karam
Department of Land Management, Faculty of Agriculture, Universiti Putra Malaysia (UPM), Serdang, Selangor, Malaysia

Z. Samdin et al. (eds.), *Tropical Forest Ecosystem Services in Improving Livelihoods For Local Communities*, https://doi.org/10.1007/978-981-19-3342-4_2

Keywords

Ecosystem services · Forest pathology · Forest pests and diseases · Pathogen · Agricultural activities · Forest sustainability · Human well-being

2.1 Introduction

Ecosystem services are defined as the benefits people obtain from ecosystems through transforming resources into a flow of essential goods and services (Price, 2007; Douglas, 2017). Forest ecosystem, for instance, is described as the community of plants, animals, microbes, and all other organisms in interaction with the chemical and physical features of their environment (Sen, 2018). Between 2001 and 2005, the ecosystem services were fully assessed by the Millennium Ecosystem Assessment (MA) (Douglas, 2017). The MA covers both unmodified (e.g., natural forests to landscapes with mixed patterns) and modified ecosystems (e.g., agricultural land and urban areas). The MA has grouped the ecosystem services into four different categories: (1) provisioning services such as food, water, timber, and fiber; (2) regulating services that affect climate, floods, disease, wastes, and water quality; (3) cultural services that provide recreational, aesthetic, and spiritual benefits; and (4) supporting services such as soil formation, photosynthesis, and nutrient cycling (Table 2.1). Thus, irrespective of culture and technology modernization, humankind depends primarily on the flow of ecosystem services.

Abundant reports have shown more and more forests across the world are affected by globalization. Expanding international trade has irrefutably increased the movement of numerous biotic stress factors such as pests, pathogens, and parasitic plants into new regions, associated with massive tree mortality and significant forest

Table 2.1 Ecosystem services are made up of provisioning, regulating, and cultural services that directly affect people and supporting services to maintain the other services

Provisioning services	Regulating services	Cultural services
Products obtained from ecosystems: ✓ Food ✓ Fresh water ✓ Fuelwood ✓ Fiber ✓ Biochemicals ✓ Genetic resources	Benefits obtained from regulation of ecosystem process: ✓ Climate regulation ✓ Disease regulation ✓ Water regulation ✓ Water purification ✓ Pollination	Nonmaterial benefits obtained from ecosystems ✓ Spiritual and religious ✓ Recreation and ecotourism ✓ Aesthetic ✓ Inspirational ✓ Educational ✓ Sense of place ✓ Cultural heritage
Supporting services		
Services necessary for the production of all other ecosystem services ✓ Soil formation ✓ Nutrient cycling ✓ Primary production		

damage, leading to economic impacts and losses of ecosystem goods and services provided by forests (Ramsfield et al., 2016). From the genetic side of views, migration causes commensal species within their original tree hosts to become a very damaging pest or pathogen due to genetic change in the infecting species. Dutch Elm disease, for instance, was developed from introduced species that acquired genes from native species that may prevent the attenuation of infection (Boyd et al., 2013). Besides, the presence of compatible mating types and sexual reproduction increases the opportunity for the newly introduced species to develop their pathogenicity to colonize native hosts. Such examples are the highly aggressive *Phytophthora* of alder that arose from a swarm of new interspecific hybrids and the spread of the aggressive *Phytophthora alni* subspecies *alni* along Bavarian rivers, which was associated with infested alder nursery stock (Cooke et al., 2000). Insect pests are less likely to hybridize and acquire new genes, but in some cases, the tree disease occurs by developing new associates, the bacterial symbionts. Likewise, China's scolytid beetle *Dendroctonus valens* acquired novel indigenous isolates of its fungal associate *Leptographium procerum*, to attract further beetles for the tree's defenses by releasing volatiles (Lu et al., 2011).

The invasive pests and pathogens are escalating along with climate change. Even though some effects of climate change are positive, it is commonly related to unfortunate events such as drought, flooding, severe gales, and high-temperature events. They are known as abiotic stress factors that increase forest disease and insect pest outbreaks and, at the same time, decrease the growth of the trees and cause massive tree mortality. Some of the most significant genera in plantation forests, such as *Pinus*, *Populus*, and *Eucalyptus*, have suffered the impact. Extreme climate change may also be the reason for the extinction of large discrete patches of trees due to forest burning and wind damage (Kurz et al., 2008). Yet, the combination of both stress factors either simultaneously or sequentially caused enormous ecological and economic importance contributing to the decline of the global forest area and biodiversity as the outcomes of local extinction of endangered species of trees, emission of biogenic volatile organic compounds, as well as a reduction in the productivity and quality of forest products and services (Teshome et al., 2020).

Much of the damage caused by tree pests and diseases befall onto provisioning services, such as timber, pulp, fruit, nuts, and vegetable oil (Ayres & Lombardero, 2000). An invasive pest may attack a tree, affect a product's availability, disturb the usual flow, and lead to a price response. The significant forest health also affects the regulating services (Trumbore et al., 2015), particularly climate regulation, carbon storage, carbon sequestration, flood control, and water purification that benefited landowner, the region, and people. Meanwhile, effects on cultural services and supporting services are evident when iconic species are severely affected by a disease and cause the amenity value of woods (Donovan et al., 2013).

To date, biological control is the best option for pests and pathogens (Kenis et al., 2017). Some researchers have suggested using microorganisms such as mycorrhiza and endophytes to alleviate abiotic stress tolerance (Teshome et al., 2020). Besides, ascertainment of the organisms of interest is a critical requirement for better disease management strategies (Ramsfield et al., 2016). Other than that, the availability of molecular tools is useful for identifying invasive organisms and further studying the populations of organisms in economically significant contexts (Adamowicz, 2015). Additional forest management strategies that can be applied to mitigate the disease impacts are thinning. For example, reduction of the basal area of stands, facilitating regeneration in advance of predicted hotter droughts, shorter rotation age to minimize damage from bark beetle and droughts, as well as stand diversification such as "clonal composites" (Teshome et al., 2020).

2.2 Ecosystem Services in Forest Pathology

Biodiversity is thought to impact a variety of services positively. Multiple ecosystem services have favorable to positive hump-shaped correlations with tree species diversity in production forests.

2.2.1 Provisioning Services

The mechanisms by which the environment provides advantages to people are known as ecosystem services. The linkages between ecosystems and human societies are known as ecosystem services. Services are generated by ecosystem components, both living and non-living, due to ecological conditions and processes. The ecosystem services framework is widely used to assess human–environment interactions, such as the MEA. Ecosystem services can be categorized into four major areas in the MEA (Balvanera et al., 2016). First is provisioning services that included food, water, fiber, fuel, and genetic resources production. Provisioning services can be defined as physical products with a well-defined market. As the population grows, so does the need for these life-sustaining services.

According to Mohammed (2019), species diversity and ecosystem service provisioning have a significant relationship in most situations, with genetic diversity being a subset of biodiversity. If the components differ in susceptibility, one of the main goals of using genetic mixes for disease control is to delay the pathogen's spread by decreasing the rate and incidence of infection. In the cases of grain rusts and powdery mildew, mixtures lowered disease severity by 40–80% (Morgounov et al., 2011). The genetic information included in particular crop types is crucial for the development of new varieties that are resistant to extreme environments such as heat, drought, salt, pests, and diseases and are fast-growing and high-yielding, which is required to address food insecurity in the face of climate change (Raza et al., 2019).

2.2.2 Supporting Services

Supporting services are ecosystem services that are required to provide all other ecosystem services. It includes natural processes such as biomass generation, atmospheric oxygen production, soil formation and retention, nutrient and water cycling, and habitat provisioning. They vary from provisioning, regulating, and cultural services in that their effects on individuals are either indirect or take a long time to manifest. In contrast, changes in the other categories have a very immediate short-term impact on people. Ecosystems could not exist without the constancy of underlying natural processes. For example, the soil is the basis of terrestrial ecosystems, and soil provides the bulk of ecosystem services required for human life (Baer & Birgé, 2018). Soil ecosystem services can be supportive of primary production and biodiversity. It can also be regulatory for erosion control, water infiltration, nutrient retention, atmospheric gas management, and insect control. The production of soil is essential to the future of terrestrial ecosystems. Microorganisms and physical mechanisms degrade organic materials into tiny particles to produce soil (Johns, 2017). Plant production may be utilized directly for food and indirectly for livestock feed, and it can also be turned into fuel and other retail goods such as lumber, paper, and textiles. As a result, many fundamental necessities for life and human comfort rely on soil and fertility, health, and management to maintain productive potential.

2.2.3 Regulating Services

Regulating services are commonly known as the benefits of managing ecological processes, such as climate regulation, water, soil erosion control, biodiversity conversation, and various human diseases (Stohlgren & Holcombe, 2013). Forest pathology includes disease and pest regulation, pollination, natural disaster mitigation, and air quality regulation. Firstly, disease and pest regulation is an intermediate ecosystem service that directly impacts human health and well-being. Besides, it has the potential to be a strong influence on controlling the provision of final ecosystem services, such as food and fiber-producing crops and livestock (Haines-Young & Potschin-Young, 2010). The pathogens and pests will be regulated or kept below harmful levels by a unique combination of biotic, abiotic, and socioeconomic factors, including predators, pathogens, competitors, and hosts for biotic, as for abiotic, such as climate and agricultural and urban land use. Lastly, socioeconomic factors include disease or pest management. The United Kingdom has fewer agricultural insect pests than continental Europe because of its moderate environment, with aphids being the most common (Parry, 2013). *Rhopalosiphum padi* transmits the virus in the south of England, whereas *Sitobion avenae* transmits it in the Midlands and north of England, resulting in severe crop losses in cereals. Predators, parasitoids, and diseases are major aphid regulators. An increase in the number of predator species improves the reduction of aphid populations and the parasitoids and hoverflies' complementing impacts on aphids. A rise in polyphagous predators like beetles and mainly spiders is related to habitat variety at the landscape scale. According to Henry et al. (2009), parasitoidism is another important agricultural pest management technique. Flies and parasitoid wasps, which lay eggs on or in a host's body, in this instance pest insects, eventually kill the hosts and reduce pest transmission.

Forest ecosystems (forests, marshes, and mangroves) can assist as the mid-components of the local water cycle. It can function as a buffer between precipitation and runoff due to its ability to capture water in the canopy and free up soil water storage via root water absorption. Moreover, forests shield soil from water and wind erosion, allowing ecosystems to efficiently manage soil loss (Artidteang et al., 2015). One of the central mechanisms behind land deterioration and desertification is soil erosion. Erosion has an impact on nutrient cycling and decreases soil fertility by reducing the pool of accessible nutrients. Soil erosion changes the structure, as well as the biological and chemical characteristics, of the soil matrix dramatically (Berhe et al., 2014). The dust and sediments that arise have effects that may be as serious or more than the productivity loss suffered on the damaged site. The soil formation function is another component of this regulating service. Soil is created by the weathering of rocks and becomes productive over time as organic matter and minerals accumulate. Given the lengthy periods of soil formation, forests play a significant role in this regard.

2.2.4 Cultural Services

Cultural ecosystem services vary from other ecosystem services in that they are primarily influenced by human experience (Vasiljevic & Gavrilovic, 2019). For example, forests are frequently seen as recreational assets in developing countries, with little chance to profit from them.

On the other hand, tourism can contribute to economic growth by providing more significant revenue, jobs, environmental conservation, and financing for national park maintenance. Certain developing countries such as Kenya already utilize natural attractions to promote tourism in protected and forest regions (Anup, 2016). According to the World Tourism Organization, sustainable tourism is defined as tourism that enhances the quality of life of local populations, offers high-quality experiences for tourists, and protects the environment on which they all rely. Ecotourism is one of the fastest-growing tourism categories globally, expanding at a rate of more than 20% yearly, two to three times faster than the total tourism business (Kiper, 2013). Several countries, including Egypt, Hungary, Laos, the Philippines, and Tunisia, have received FAO technical support to develop ecotourism based on sustainable forest use.

Aesthetics value is frequently mentioned as a cultural service example. The way the forest is managed impacts its physical appearance and, as a result, public perception (Wohlwill, 1976). Visual appeal is always linked to environmental sustainability in the eyes of the general community. As a result, the good visual quality is critical for the societal perception of management methods, which is necessary for long-term sustainability. Therefore, improving the visual quality of forests would significantly impact human mental and physical health, pleasant aesthetic experience, and overall well-being (Franco et al., 2017).

2.3 Impact of Forest Diseases on Ecosystem Services in Malaysia

One of the main challenges in the forest ecosystem is managing the impact of the forest disease itself. Before 1970, there are very few diseases founded in the Malaysian forest plantation. Then, however, various prominent diseases started to emerge and affect timber quality and production, including heart rot disease, red root rot caused by *Ganoderma philippii*, and vascular wilt disease due to *Ceratocystis* spp. (Ratnasingam et al., 2020). The main impact of forest disease in provisioning services is translated to crop yield (Cheatham et al., 2009). However, due to multiple diseases in Malaysian forest plantations that have no effective management, some species were abandoned in favor of fast-growing tree species such as rubber trees (*Hevea brasiliensis*). The scenario can be seen in *Acacia mangium*, which was discarded due to its susceptibility to heart rot disease (Ratnasingam et al., 2020).

On a bigger scale of the impact of forest disease, it may affect regulating services such as climate and water regulation and water purification. These services provided by the plant could not be replaced artificially and especially forest plant that acts as a carbon sink. The canopy provided by the forest plant slows down rain velocity, and the root secures soil and prevents landslides (Cheatham et al., 2009). Surface soil erosion (Fig. 2.1) and runoff (Sidle et al., 2006) could be very prominent due to forest plant removal.

Fig. 2.1 Surface soil erosion whereby water and sediment are discharged at the node connected to the headwater stream (Sidle et al. 2006)

Carbon cycling is also significantly impacted by forest disease, whereby fungal infection on the plant could reduce the photosynthesis rate of a plant. This, in turn, will affect overall provisioning service and regulation services provided by forest (Paseka et al., 2020). In addition, supporting services provided by forests, especially on nutrient cycling and soil formation, will also be impacted. For example, in Malaysia, plantation forest was established on soil with poor fertility, and fertilization was only carried out at the initial tree growth (Abd Latif et al., 2018).

Disease management plays a vital role in establishing a healthy forest for the ecosystem and services. While disease management is essential, it could also impact the ecosystem if mismanaged. For example, a large scale of forest removal could cause weaker methane (CH_4) uptake and greater nitrogen oxide (N_2O) emission, which can accelerate global warming (Yashiro et al., 2008).

2.4 Impact of Agricultural Activities

Forest trees are vulnerable to a broad range of pests and diseases caused by viruses, fungi, bacteria, oomycetes, nematodes, and insects, which may inflict catastrophic damage to forest natural resources, resulting in significant changes in forest biodiversity and the structure and functions of forest ecosystem services. Forest diseases are responsible for tree deaths and are also responsible for decreases in tree growth and tree growth anomalies. Initially, damages begin with seed abortion and a reduction in seed germinability and seedling survival. Disease symptoms typically emerge due to a complicated interplay between the vulnerable host, predisposing environmental circumstances, vectors, and a living infectious agent such as an insect. Tree disease causes are often classified as either non-infectious or infectious (Parthasarathy et al., 2021). A recent assessment has determined that output in forests of Malaysia is below potential since resource consumption is proliferating and demand increases faster than supply (Food and Agriculture Organization of the United Nations, 2020).

Acacia mangium (*A. mangium*) plantations are dedicated to manufacturing general-purpose timber, while East Malaysia is dedicated to the pulp and paper sector. Two significant diseases of *A. mangium* in Malaysia, heart rot and root rot, have undoubtedly caused more timber mortality than any other forest disease in Malaysia (Lee, 2004). White fibrous rot is the most frequent kind of heart rot disease observed in *A. mangium*, and it can appear in tiny pockets of the heartwood, resulting in reduced timber yield, especially on small logs (Tarigan et al., 2011). The decay is only visible after the tree is cut down. The incidence of heart rot in *A. mangium* is very high, around 50–98%, but the volume of wood affected is less than 10% (Lee, 2004). Because of structural strength loss, inappropriate stain, and inadequate recovery of high-value products during processing, timber with heart rot disease is not appropriate for general uses such as constructing and appearance grades. Another disease is root rot caused by *Ganoderma philippii*, which will significantly impact *A. mangium* plantations. The leaves of the trees infected are light green and are decreased in size and quantity (Ratnasingam et al., 2020). A long-term root disease study in Malaysia found more than 40% mortality in highly infected regions in *A. mangium* plantations in trees aged 10–14 years (Ratnasingam et al., 2020).

Gypsy moth *Lymantria dispar* L. was introduced in Massachusetts in 1869 to develop a resistant silkworm. However, gypsy moth epidemics cause significant defoliation in Holarctic woods in North America, where it is invasive, and Eurasia, where it is native (Alalouni et al., 2013). Defoliation of forests and urban trees may severely impact individual trees, ecosystem components, and humans and hinder timber production and the natural population (Leroy et al., 2021). Several significant outbreaks have been recorded in its native range over several decades, covering much of Europe, North Africa, Central Asia, and Japan (Alalouni et al., 2013; Lentini et al., 2020). According to recent estimates, the total financial toll of the gypsy moth invasion in North America is a staggering US$3.2 billion per year, including government expenses and timber losses. Up to 90% tree mortality often occurs 2–3 years following a severe defoliation incident. Also, it can be caused by secondary parasites such as *Armillaria mellea* and *Agrilus bilineatus*, which readily attack severely damaged trees (Mcmanus & Csóka, 2007).

Sirex woodwasp disease killed about 30% of the radiata pine tree population with a range of 120,000 ha plantation in New Zealand. It killed 40% of the pine tree population, with a range of 1092 ha in Tasmania (Ayres et al., 2014). The Sirex woodwasp epidemic in Australia has killed five million trees worth AUS $10–$12 million. The larvae's tunneling activity destroys the wood, and if they dig the wood for half to a year, the timber becomes unsuitable (Ayres et al., 2014). The occurrence and prevalence of forest pests and diseases have significantly impacted a country's revenue and resources, especially in timber production due to tree mortality (Table 2.2).

2.5 Conclusion

Forest ecosystems contribute to the nonmaterial advantages that result from people's interactions with ecosystems through cultural services. It contributes significantly to human well-being by offering societal regulating services. The cultural services would also be impacted due to forest disease. As a country known for its rich biodiversity, Malaysia relies very much on our nature for ecotourism. Based on current trends in pest and disease outbreaks, there are urgent needs for prevention strategies and the required method to protect the forest. Thus, the forest management and government need to control and take countermeasures to prevent forest diseases efficiently.

Table 2.2 Example of forest pests and diseases along with their impacts

Pest or disease	Host	Native range	Damage and impacts
Sal heartwood borer	*Hoplocerambyx spinicornis*, commonly known as Sal borer (beetle)	Indian subcontinent, Indonesia, Malaysia, and Philippines	Extensive damage to Sal trees both in standing and freshly felled timber. Weak and unhealthy trees are easily attacked and killed by the borer
	Shorea robusta (Sal tree)		Epidemic of Sal borer have been reported from the forest of different country with tree death more than 1% of the total number of trees before matures lead to economic losses in those countries
Emerald ash borer	*Agrilus planipennis* (beetle)	China, Korea, Japan, and Russian far east	Larvae infestation producing yellowing and thinning leaves as well as tree dieback and mortality within 3 years
	Ash, Japanese wingnut, walnut and elm tree		Millions of ash trees in Canada, the United States, and Moscow forests have died or declined due to the infestation. Because of the fast spread of the pest, European woodlands are also in danger of being infested
Chestnut blight disease	*Cryphonectria parasitica* (fungi)	Asia	Infects tree branches and trunks above ground, generating cankers that spread, girdle, and finally kill the tree branches and trunks
	Castanea dentata (chestnut tree) and *Quercus* spp. (oak)		The American chestnut was formerly one of the most common hardwoods in the eastern United States. Still, it is now pretty much extinct due to chestnut blight disease, demonstrating how a condition can profoundly change an entire ecosystem
Dutch elm disease	*Ophiostoma ulmi* and *Ophiostoma novo-ulmi* (fungi)	Asia	The fungus spreads by tree sap and may also apply through root grafts from tree to tree. Wilting, yellowing, browning of leaves, discoloration in branches and stems

(continued)

Table 2.2 (continued)

Pest or disease	Host	Native range	Damage and impacts
	Ulmus spp. (elm tree)		Hundreds of millions of healthy adult elms were destroyed in northern Asia, Europe, North America, and France, resulting in the decimated supply of American elm lumber and veneer
Pine wilt disease	*Bursaphelenchus xylophilus* (nematode)	North America	Nematodes in the xylem can cause wilt and death, and they also feed on fungal tissues in dead trees or wood products
	Pinus spp. (pine tree)		Threat to some pine woods and resulted in widespread tree mortality in some regions, with millions of trees lost each year in Japan
Pink disease	*Erythricium salmonicolor* (fungi)	Africa, Asia, Europe, North America, Oceania, and South America	The symptoms begin with swelling or sunken regions on the main stem and branches, followed by cracking or splitting bark and stem canker formation
	Teak wood (*Tectona grandis*), cocoa, rubber, African mahogany, and tropical woody tree		Pink plague can result in substantial losses, ranging from the death of individual branches to the death of the entire tree if the main stem or numerous branches are afflicted
Heart rot disease	*Polyporus* spp. and *Phellinus* spp. (fungi)	Asia	Conks or mushrooms are frequently found near wounds or at the trunk's base. Heart rot in trees does not generally harm the live sapwood, but it may cause structural flaws and result in broken branches and trunks
	Acacia mangium, oaks and most of hardwood tree		The disease generates economic losses in the lumber sector because infected trees are frequently unsuitable for timber production

(continued)

Table 2.2 (continued)

Pest or disease	Host	Native range	Damage and impacts
Sirex woodwasp	*Sirex noctilio* (wasp)	Europe, Asia, and North Africa	*Sirex noctilio* secretes a toxic mucus and a symbiotic fungus into the tree, both of which work together to weaken and, in some circumstances, kill the infested tree
	Pinus radiata, *Pinus taeda*, and other *Pinus* tree		Millions of North American pines planted in Southern Hemisphere woods have died due to the Sirex woodwasp disease. The illness caused by the *Sirex noctilio* is currently expected to cost between $16 and $60 million each year
Bacterial wetwood (slime flux)	*Klebsiella* spp., *Clostridium* spp., and *Pseudomonas* spp. (bacteria)	United States	It affects the center core of many shades and forest trees, resulting in bleeding on trunks and branches and a crack and a foul stench of slimy liquid produced
	Pinus spp., *Abies* spp., *Acer* spp., *Quercus* spp., *Populus* spp., and *Ulmus* spp.		The liquid damages grass and other herbaceous plants that come into touch with it near the tree's base. Because of the unattractive discoloration, the wood of afflicted trees has a significantly reduced value as timber

References

Abd Latif, M., Wan Rasidah, K., & Ahmad Zuhaidi, Y. (2018). Is plantation forestry a wise investment? A perspective from Malaysia's initiatives. *Journal of Tropical Forest Science, 30*(5), 461–467. https://doi.org/10.26525/jtfs2018.30.5.461467

Adamowicz, S. J. (2015). International Barcode of Life: Evolution of a global research community. *Genome, 58*, 151–162. https://doi.org/10.1139/gen-2015-0094

Alalouni, U., Schädler, M., & Brandl, R. (2013). Natural enemies and environmental factors affecting the population dynamics of the gypsy moth. *Journal of Applied Entomology, 137*(10), 721–738.

Anup, C., Jitendra, K., Raghuveer, S., Bharat, S., Bhoben, P., & Rupankar, B. (2016). Drenching treatment for the rectification of trunk-borer, Anoplophora versteegi Ritsema (Coleoptera; Cerambycidae) infested citrus plants. *Environment and Ecology, 34*(3B), 1273–1276.

Artidteang, S., Bergado, D. T., Chaiyaput, S., & Tanchaisawat, T. (2015). Embankment reinforced with limited life geotextiles on soft clay. *Proceedings of the Institution of Civil Engineers-Ground Improvement, 168*(2), 130–143.

Ayres, M. P., & Lombardero, M. J. (2000). Assessing the consequences of global change for forest disturbance from herbivores and pathogens. *Science of the Total Environment, 262*, 263–286. https://doi.org/10.1016/S0048-9697(00)00528-3

Ayres, M. P., Pena, R., Lombardo, J. A., & Lombardero, M. J. (2014). Host use patterns by the European Woodwasp, *Sirex noctilio*, in its native and invaded range. *PLoS One, 9*(3), e90321.

Baer, S., & Birgé, H. (2018). *Soil ecosystem services: An overview*. Burleigh Dodds Science Publishing. https://doi.org/10.19103/AS.2017.0033.02

Balvanera, P., Quijas, S., Martín-López, B., Barrios, E., Dee, L., Isbell, F., Durance, I., White, P., Blanchard, R., & de Groot, R. (2016). The links between biodiversity and ecosystem services. In *Routledge handbook of ecosystem services* (pp. 45–61). Routledge.

Berhe, A. A., Arnold, C., Stacy, E., Lever, R., McCorkle, E., & Araya, S. N. (2014). Soil erosion controls on biogeochemical cycling of carbon and nitrogen. *Nature Education Knowledge, 5*(8), 2.

Boyd, I. L., Freer-Smith, P. H., Gilligan, C. A., et al. (2013). The consequence of tree pests and diseases for ecosystem services. *Science, 342*, 1235773. https://doi.org/10.1126/science.1235773

Cheatham, M. R., Rouse, M. N., Esker, P. D., et al. (2009). Beyond yield: Plant disease in the context of ecosystem services. *Phytopathology, 99*(11), 1228–1236. https://doi.org/10.1094/PHYTO-99-11-1228

Cooke, D. E. L., Drenth, A., Duncan, J. M., et al. (2000). A molecular phylogeny of Phytophthora and related oomycetes. *Fungal Genetics and Biology, 30*, 17–32. https://doi.org/10.1006/fgbi.2000.1202

Donovan, G. H., Butry, D. T., Michael, Y. L., et al. (2013). The relationship between trees and human health: Evidence from the spread of the emerald ash borer. *American Journal of Preventive Medicine, 44*, 139–145. https://doi.org/10.1016/j.amepre.2012.09.066

Douglas, I. (2017). Ecosystems and human well-being. In *Encyclopedia of the anthropocene* (Vol. 1–5, pp. 185–197). Elsevier. https://doi.org/10.1016/B978-0-12-809665-9.09206-5

Food and Agriculture Organization of the United Nations. (2020). *Global forest resources assessment 2020: Main report*. Food and Agriculture Organization of the United Nations.

Franco, L. S., Shanahan, D. F., & Fuller, R. A. (2017). A review of the benefits of nature experiences: More than meets the eye. *International Journal of Environmental Research and Public Health, 14*(8), 864. https://doi.org/10.3390/ijerph14080864

Haines-Young, R., & Potschin-Young, M. (2010). The links between biodiversity, ecosystem service and human well-being. *Ecosystem Ecology: A New Synthesis, 1*, 110–139. https://doi.org/10.1017/CBO9780511750458.007

Henry, L. M., Ma, B. O., & Roitberg, B. D. (2009). Size-mediated adaptive foraging: A host-selection strategy for insect parasitoids. *Oecologia, 161*(2), 433–445.

Johns, C. (2017). *Living soils: The role of microorganisms in soil health—Future directions international*. Retrieved August 25, 2021, from https://www.futuredirections.org.au/publication/living-soils-role

Kenis, M., Hurley, B. P., Hajek, A. E., et al. (2017). Classical biological control of insect pests of trees: Facts and figures. *Biological Invasions, 19*, 3401–3417. https://doi.org/10.1007/s10530-017-1414-4

Kiper, T. (2013). Role of ecotourism in sustainable development. In *Advances in landscape architecture*. IntechOpen.

Kurz, W. A., Dymond, C. C., Stinson, G., et al. (2008). Mountain pine beetle and forest carbon feedback to climate change. *Nature, 452*, 987–990. https://doi.org/10.1038/nature06777

Lee, S. S. (2004). Diseases and potential threats to *Acacia mangium* plantations in Malaysia. *Unasylva (English ed.), 55*(217), 31–35.

Lentini, A., Mannu, R., Cocco, A., Ruio, P. A., Carboneschi, A., & Luciano, P. (2020). Long-term monitoring and microbiological control programs against lepidopteran defoliators in Sardinian cork oak forests (Italy). *Annals of Silvicultural Research, 45*(1), 21–30.

Leroy, B. M. L., Lemme, H., Braumiller, P., et al. (2021). Relative impacts of gypsy moth outbreaks and insecticide treatments on forest resources and ecosystems: An experimental approach. *Ecological Solutions and Evidence, 2*, e12045.

Lu, M., Wingfield, M. J., Gillette, N., et al. (2011). Do novel genotypes drive the success of an invasive bark beetle–fungus complex? Implications for potential reinvasion. *Ecology, 92*, 2013–2019. https://doi.org/10.1890/11-0687.1

Mcmanus, M., & Csóka, G. (2007). History and impact of gypsy moth in North America and comparison to recent outbreaks in Europe. *Acta Silvatica et Lignaria Hungarica, 3*, 47–64.

Mohammed, J. (2019). The role of genetic diversity to enhance ecosystem service. *American Journal of Biological and Environmental Statistics, 5*(3), 46–51.

Morgounov, A., Tufan, H., Ram, S., Akin, B., Bağcı, S., Braun, H.-J., Kaya, Y., Keser, M., Payne, T., Sonder, K., & Mcintosh, R. (2011). Global incidence of wheat rusts and powdery mildew during 1969–2010 and durability of resistance of winter wheat variety Bezostaya 1. *European Journal of Plant Pathology, 132*(3), 323–340. https://doi.org/10.1007/s10658-011-9879-y

Parry, H. R. (2013). Cereal aphid movement: General principles and simulation modelling. *Movement Ecology, 1*, 14. https://doi.org/10.1186/2051-3933-1-14

Parthasarathy, S., Thiribhuvanamala, G., Muthulakshmi, P., & Angappan, K. (2021). *Diseases of forest trees and their management* (1st ed.). CRC Press.

Paseka, R. E., White, L. A., Van de Waal, D. B., et al. (2020). Disease-mediated ecosystem services: Pathogens, plants, and people. *Trends in Ecology & Evolution, 35*(8), 731–743. https://doi.org/10.1016/j.tree.2020.04.003

Price, R. (2007). *An introductory guide to valuing ecosystem services*. Department of Environment, Food and Rural Affairs.

Ramsfield, T. D., Bentz, B. J., Faccoli, M., et al. (2016). Forest health in a changing world: Effects of globalization and climate change on forest insect and pathogen impacts. *Forestry, 89*, 245–252. https://doi.org/10.1093/forestry/cpw018

Ratnasingam, J., Latib, H., Paramjothy, N., Liat, L., Nadarajah, M., & Ioras, F. (2020). *Plantation forestry in Malaysia: An evaluation of its successes and failures since the, 1970*, 1789–1801.

Raza, A., Razzaq, A., Mehmood, S. S., Zou, X., Zhang, X., Lv, Y., & Xu, J. (2019). Impact of climate change on crops adaptation and strategies to tackle its outcome: A review. *Plants (Basel, Switzerland), 8*(2), 34. https://doi.org/10.3390/plants8020034

Sen, D. (2018). *Information about a forest ecosystem*. Retrieved from https://sciencing.com/about-6325114-information-forest-ecosystem.html

Sidle, R. C., Ziegler, A. D., Negishi, J. N., et al. (2006). Erosion processes in steep terrain—Truths, myths, and uncertainties related to forest management in Southeast Asia. *Forest Ecology and Management, 224*(1–2), 199–225. https://doi.org/10.1016/j.foreco.2005.12.019

Stohlgren, T. J., & Holcombe, T. (2013). Impacts of land-use change to ecosystem services. In *Climate vulnerability* (pp. 13–22). Springer.

Tarigan, M., Roux, J., Van Wyk, M., Tjahjono, B., & Wingfield, M. J. (2011). A new wilt and dieback disease of *Acacia mangium* associated with *Ceratocystis manginecans* and *C. acaciivora* sp. *nov*. in Indonesia. *South African Journal of Botany, 77*(2), 292–304.

Teshome, D. T., Zharare, G. E., & Naidoo, S. (2020). The threat of the combined effect of biotic and abiotic stress factors in forestry under a changing climate. *Frontiers in Plant Science, 11*, 601009. https://doi.org/10.3389/fpls.2020.601009

Trumbore, S., Brando, P., & Hartmann, H. (2015). Forest health and global change. *Science, 349*, 814–818. https://doi.org/10.1126/science.aac6759

Vasiljevic, N., & Gavrilovic, S. (2019). Cultural ecosystem services. In F. W. Leal, A. Azul, L. Brandli, P. Özuyar, & T. Wall (Eds.), *Life on land. Encyclopedia of the UN sustainable development goals*. Springer.

Wohlwill, J. F. (1976). Environmental aesthetics: The environment as a source of affect. In *Human behavior and environment* (pp. 37–86). Springer.

Yashiro, Y., Kadir, W. R., Okuda, T., et al. (2008). The effects of logging on soil greenhouse gas (CO_2, CH_4, N_2O) flux in a tropical rain forest, Peninsular Malaysia. *Agricultural and Forest Meteorology, 148*(5), 799–806. https://doi.org/10.1016/j.agrformet.2008.01.010

Non-timber Forest Problems: NTFPs in Conservation and Development Initiatives

3

Melaina Dyck

Abstract

Non-timber forest products (NTFPs) are any goods and services derived from forests besides timber. NTFPs have been consumed, traded and managed by forest-dependent peoples for millennia. Starting in the 1980s, NTFPs became the focus of a variety of conservation and development programmes that viewed commercialisation of NTFPs as a strategy to incentivise conservation of forests by improving forest community livelihoods. However, these initiatives often overlook existing local systems that promote sustainable forest management, leading to marginalisation of forest-dependent communities and overexploitation of NTFP species. To be effective, NTFP-based initiatives must understand NTFP species as cultural-ecological keystones in the management approaches of forest-dependent peoples and promote secure land rights that enable community-led conservation.

Keywords

NTFP (non-timber forest product) · Forest-dependent · Communities · Marginalisation · Overexploitation · Commercialisation · Conservation · Land rights · Amazon

M. Dyck (✉)
Independent Scholar, Chapel Hill, North Carolina, USA
e-mail: melaina.dyck@aya.yale.edu

Z. Samdin et al. (eds.), *Tropical Forest Ecosystem Services in Improving Livelihoods For Local Communities*, https://doi.org/10.1007/978-981-19-3342-4_3

3.1 Introduction

Non-timber forest products (NTFPs) have been an integral part of human societies for millennia. NTFPs are collected for subsistence, trade and cultural practices by millions of forest-dependent people (Girão Rodrigues de Mello et al., 2020; Ros-Tonen, 2000; Shackleton et al., 2018). At least 150 NTFPs are traded in international commodity markets (Girão Rodrigues de Mello et al., 2020; Ros-Tonen, 2000). Over the past three decades, NTFPs have been at the centre of numerous initiatives to conserve forests while providing sustainable development to forest-dependent communities. However, such programmes have often failed to meet economic, ecological or social objectives, in part due to disregard for power disparities and the existing socio-cultural roles of NTFPs (Girão Rodrigues de Mello et al., 2020). NTFPs need to be analysed and developed within socio-ecological systems (Girão Rodrigues de Mello et al., 2020; Shackleton et al., 2018).

This chapter examines the use of NTFP initiatives for conservation and development in tropical forest regions. The chapter traces the emergence of NTFPs as an area of inquiry and unpacks the assumptions that underlie NTFP-based initiatives to explore why NTFP-based initiatives do not achieve their goals. The chapter concludes with a discussion of the need for community-led management of NTFP and forest resources. Although NTFPs are derived from forest ecosystems all over the world, they are most abundant in and most frequently associated with tropical forests. The foundational study that brought NTFPs to the attention of academic and rural development fields was based on an economic valuation of NTFPs in the Peruvian Amazon (Peters et al., 1989). As such, the examples in this paper pertain to tropical forests and the Amazon in particular.

3.2 What Are NTFPs?

By the broadest definition, "non-timber forest products" are any items and services derived from forest ecosystems apart from trees logged for timber. NTFPs include resources such as fruits, tubers, meats, forage, fuels, medicines, fibres, biochemicals, fur, feathers, construction materials, rubber, silk, gums, resins, seeds, oils, dyes, tannins and inorganic materials as well as ecosystem services like clean water, natural cooling or ecotourism attractions (Aguirre Mendoza & Aguirre Mendoza, 2021; Belcher, 2003; FAO & FILAC, 2021; Girão Rodrigues de Mello et al., 2020; Padoch, 1988; Ros-Tonen, 2000; Shackleton & Shackleton, 2004). The harvest of NTFPs from unmanaged and managed forest environments has been essential to forest-dependent societies for millennia (Aguirre Mendoza & Aguirre Mendoza, 2021). For forest-dependent communities, NTFPs may form the primary basis for subsistence consumption and livelihoods (Rist et al., 2012), provide periodic supplemental income (Penn, 2008), fill subsistence gaps in lean times (Belcher et al., 2005) or raise money for special purchases (Dove & Kammen, 1997).

The term "non-timber forest product" was introduced in 1989 as an alternative to the previously accepted moniker, "minor" forest products, reflecting an emerging understanding that "minor" forest products play a *major* role in forest-based economies, consumption patterns and cultures (Belcher, 2003; de Beer & McDermott, 1996; Padoch, 1988; Peters et al., 1989). Interest in NTFPs aligned with a political shift in conservation and development programmes to focus on environmental sustainability and the rights of indigenous peoples, particularly in tropical forest regions (Belcher et al., 2005; Girão Rodrigues de Mello et al., 2020). Initiatives were developed to stimulate the commercialisation of NTFPs as a strategy to incentivise forest conservation through sustainable development (Aguirre Mendoza & Aguirre Mendoza, 2021; Blaser et al., 2021; Girão Rodrigues de Mello et al., 2020). Conservation and development programmes positioned NTFP commercialisation as a mechanism to advance conservation and economic development, particularly in the rural tropics for forest-dependent people—those who live in and near forests and rely on forests for their livelihoods (FAO & FILAC, 2021; Ros-Tonen & Wiersum, 2005). The phrase "NTFP" thus carries connotations of "aid for poor people" and "incentivizing conservation" in tropical forest regions (Belcher, 2003).

NTFPs may come from managed forest and shifting cultivation or swidden-fallow[1] agriculture systems as well as unmanaged ecosystems. Many NTFPs are undergoing a transition from wild foraged products to domesticated in gardens, swiddens or agroforestry systems (Blaser et al., 2021; Ros-Tonen, 2000). This is the case in the Amazon, for example, where landscapes are undergoing a steady change from primary forest to cropland, orchard, fallow and secondary forest (Coomes et al., 2016). As landscapes change, patterns of extraction and production shift, and defining NTFPs based on how they are derived becomes challenging. Which resources "count" as NTFPs is contested because no single definition encapsulates the diverse cultural, ecological, subsistence and economic roles and processes of these varied products (Belcher, 2003; Belcher et al., 2005; Boon & Ahenkan, 2011).

In the context of this literature review, NTFPs are understood as ecosystem goods *and* services that are central to the economies and cultures of forest-dependent peoples and conservation and development projects. This definition identifies NTFPs based on their roles rather than the ecosystems they come from and situates NTFPs within the economic-political context of conservation and development programmes. This chapter also calls for NTFPs to be understood as objects of knowledge and innovation by forest-dependent people, who often have the first and greatest understanding of NTFP management, uses and extraction (Drahos & Frankel, 2012; FAO & FILAC, 2021; Penn, 2008).

[1] Swidden-fallow cultivation encompasses a variety of agricultural systems that involve clearing, cultivating and fallowing of forest landscapes. Often, swidden cultivators farm the same territory over multi-year cyclical rotations, cultivating sections of the landscape for a short time and then allowing forest to regrow in those areas for a longer time while cultivating other sections (Dove, 2011).

3.3 Commercialisation for Conservation and Development

NTFPs are the focus of a wide range of forest conservation and sustainable development initiatives in the tropics because they are positioned as "benign" commodities that can be extracted by forest-dependent communities without degradation of forest ecosystems (Blaser et al., 2021; Ros-Tonen, 2000). Conservation and development projects aim to find a commercially valuable NTFP that is undeveloped or unknown, advertise to create a market, encourage forest-dependent communities to initiate or increase extraction of the product, send the product processed or raw to the market—sometimes via intermediate traders—and generate sufficient income for forest-dependent communities to avoid deforestation (Ahenkan & Boon, 2010).

NTFP initiatives assume that although forest-dependent communities use NTFPs, communities do not understand the value of NTFPs, have not developed their NTFP resources and have no sources of income other than those that drive deforestation (Belcher et al., 2005; Dove, 1993, 1994). Initiatives also assume that forest-dependent people are the primary drivers of deforestation and thus, with sufficient financial offers, they could stop deforestation (Blaser et al., 2021; Dove, 1994; Girão Rodrigues de Mello et al., 2020). Following these assumptions, initiatives to commercialise NTFPs theorise that creating market value for NTFPs will incentivise sustainable forest management (Blaser et al., 2021; Ros-Tonen, 2000).

Green certification programmes and green marketing campaigns exemplify how the assumptions behind NTFP-based initiatives are applied. Certification schemes for NTFPs are premised on the notion that consumers in developed countries will pay more for products that are harvested or produced "sustainably", thus incentivising poor people in tropical, developing countries to conserve and sustainably use their forests (Dove, 1994; Quaedvlieg et al., 2014). This places responsibility for tropical deforestation on individual consumers and on forest-dwelling communities who are supposed to respond to changes in market demand even while their own needs for and knowledge of the forest are ignored (Dove, 1994). Certification programmes do not address the systemic problems that threaten both forest ecosystems and forest-dwelling communities (Dove, 1994)—namely, that forest resources are not controlled by the rural people green certification claims to incentivise, but rather by large private and state entities that continue to deforest even as they profit from higher prices associated with NTFPs they extract.

The example of green certification schemes typifies the problems with using NTFPs to drive conservation and sustainable development. The literature examined in this chapter challenges the assumptions underlying NTFP-based initiatives by showing that the marginal position of forest-dependent communities inhibits their ability to participate in NTFP markets and prevent deforestation, while NTFP commercialisation further disregards the knowledge of forest communities, drives overexploitation of NTFPs and fails to secure economic returns.

3.3.1 Marginalisation

NTFP conservation and sustainable development initiatives recognise that forest-dependent communities use and rely on forests but assume that NTFPs are undeveloped and undervalued. Such initiatives fail to recognise existing local systems of NTFP trade and may far overestimate the economic potential for NTFP-based livelihoods (Aguirre Mendoza & Aguirre Mendoza, 2021; Dove, 1993; Girão Rodrigues de Mello et al., 2020). Most NTFPs are consumed or traded locally by the people who collect them and are embedded in local economies, cultural practices and long-established systems of extraction, management and trade (Aguirre Mendoza & Aguirre Mendoza, 2021; Belcher et al., 2005). NTFPs rarely serve as the sole basis for livelihoods, but function as a diversification strategy for subsistence and income (Aguirre Mendoza & Aguirre Mendoza, 2021; Pattanayak & Sills, 2001; Ros-Tonen, 2000).

The commercialisation of NTFPs risks creating a boom–bust market with detrimental consequences for extractor communities (Osborne, 2015; Shanley et al., 2002). Rather than consuming and trading a range of NTFPs as part of subsistence and income-generating activities, rapid commercialisation may drive extractors to harvest or cultivate one product for cash income and then leave them with no sources of income when demand dries up (Li, 2014). Communities may lose access to NTFP resources altogether when commercialisation disrupts local markets and use rights. Local consumers may no longer be able to afford the product or access the species due to scarcity or loss of forest territory (Li, 2014; Osborne, 2015). Furthermore, when forest-dependent communities are successful in commercialising NTFPs, the production of and rights to forest resources are frequently taken over by outside interests. Communities may be wary of participating in commercialisation schemes, knowing that if products become valuable, they risk being appropriated as other community-managed resources have been (Dove, 1993).

Imbalanced power arrangements impact NTFP commercialisation (Girão Rodrigues de Mello et al., 2020). Unequal distribution of profits or access, lack of political support and limited market access can further marginalise forest-dwelling communities rather than provide economic opportunities (Girão Rodrigues de Mello et al., 2020). When NGOs and development agencies position NTFP commercialisation as a mechanism to incentivise reduced forest degradation, they consistently fail to distinguish between communities who subsist on NTFPs and those who possess resources to engage in NTFP markets (Aguirre Mendoza & Aguirre Mendoza, 2021; Arnold & Pérez, 2001; Girão Rodrigues de Mello et al., 2020). Control over NTFP extraction, production and markets is concentrated among wealthier, better-connected households and communities (Ambrose-Oji, 2003; Girão Rodrigues de Mello et al., 2020; Li, 2014). NTFP commercialisation risks increasing inequalities in income and resource access. Wealthier households have more time and capital to invest in NTFP cultivation and can take control of production and trade at the expense of poorer households (Ambrose-Oji, 2003; Arnold & Pérez, 2001; Coomes et al., 2016; Li, 2014). This fosters a feedback loop in which households with more NTFP assets take more risks and expand production because they have the insurance of sufficient product quantities (Coomes et al., 2016; Li, 2014). Outcomes such as these can be mitigated by strong communal management systems (Coomes, 2004), but may also introduce privatisation of land or resources that undermine communal management, further exacerbating wealth inequalities (Li, 2014; Osborne, 2015).

For example, in Ecuador, the cultivation of palm was most common among wealthier households, who had more access to land, labour and time required for palm cultivation (Byg & Balslev, 2006). In another study of Amazonian palms, overharvest of palm fibres accelerated when households perceived the palms becoming scarce. The wild stocks were depleted, and ultimately wealthier households cultivated palms on their lands while poorer households lost access to fibres and could no longer produce the handicrafts they used as a source of income (Coomes, 2004). NTFPs can be understood like other economic assets. Possessing more assets allows extractors to take risks and expand production because they have the insurance of sufficient quantities of NTFPs. Ultimately, income and security from NTFP resources can be used to access more NTFPs and agricultural resources (Coomes et al., 2016).

In addition to wealthier households receiving more of the benefits of NTFP commercialisation than poor households, the perception that NTFPs are undeveloped objects that are used by only the poorest people justifies systematic devaluation of NTFPs in global markets (Aguirre Mendoza & Aguirre Mendoza, 2021; Arnold, 2000; Dove, 1993). State and corporate elites in less-wealthy countries lack the political capital to advocate for fair prices for NTFPs in global markets but are themselves still benefitting from NTFP trade. Governments and corporations do not ensure that the benefits of NTFP trade reach extractors and producers. Systematic devaluation perpetuates the idea that NTFPs are low-value products, which undermines development goals (Dove, 1994).

Furthermore, government policies may harness conservation efforts to constrain rather than facilitate community access to forest areas, such as by agreeing that communities can harvest certain NTFPs but disallowing traditional forest management or cultural practices (Arnold & Pérez, 2001; Woods & Naimark, 2020). For example, swidden cultivation, practised by forest-dependent communities all over the world, is the only sustainable system of agriculture ever developed for the tropics (Dove, 1994). Swidden requires fewer inputs and less environmental disruption than green revolution technologies and supports one billion people worldwide (Dove & Kammen, 1997). Yet, swidden is consistently vilified by conservationists as destructive to tropical forests and governments seek to stop and regulate swidden (Arnold, 2000; Dove, 1993; Woods & Naimark, 2020).

By neglecting to consult or actively excluding forest-dependent communities, NTFP conservation and development initiatives ignore local peoples' expert understandings of how local ecosystems function. Initiatives then become irrelevant to forest-dependent people who have their own management systems in place (Branford, 2020; Dove, 1993; Li, 2007). An example from the Amazon highlights the marginalisation of NTFP harvesters. Government agencies and NGOs were promoting forest plantations in parts of the Amazon based on the theory that plantations are more profitable and environmentally sustainable than cash crop plantations or extraction of forest products (Hoch et al., 2012). However, Amazonian smallholder agroforestry systems were found to generate at least as much income as the plantations while being more resilient to environmental shocks and requiring less labour. Agroforestry was more attractive to local communities, who already engaged in agroforestry practices and faced barriers to successful plantation cultivation, including natural disasters, disease, lack of technology and limited land rights (Hoch et al., 2012). Yet, plantations were promoted, and only 1% of smallholders met the programme goals for production and commercialisation (Hoch et al., 2012).

Where NTFP markets are accessible, NTFP extraction can be an attractive income diversification strategy that offers a fairly high return for labour input (Hoch et al., 2012). However, there are a host of barriers to engaging in markets, including difficulty finding transportation over long distances from remote regions; adverse weather events that disrupt supply chains and may damage the products; perishable products that lose value quickly; lack of market experience and financial or market literacy in producer communities; and exchange intermediaries who do not pay a fair price for goods (Blaser et al., 2021; Li, 2007; Shanley et al., 2002). Social marginalisation is also a significant barrier (Aguirre Mendoza & Aguirre Mendoza, 2021). Extractors—even wealthy households—are often from socially marginalised communities (FAO & FILAC, 2021; Ros-Tonen, 2000). The lack of protection and respect for these communities means that extractors struggle to overcome commercialisation barriers, are readily exploited by intermediaries and often end up in debt to make ends meet (Li, 2014; Ros-Tonen, 2000).

3.3.2 Overexploitation

Underlying NTFP-based projects is the assumption that poor, rural people in tropical countries are mismanaging their resources and must cut down their forests unless rich countries and consumers offer financial incentives (Blaser et al., 2021; Dove, 1994). Yet, the majority of Earth's remaining tropical forests are inhabited by indigenous and smallholder communities, who have sophisticated methods to manage forest ecosystems that promote forest growth (Branford, 2020; FAO & FILAC, 2021; Shackleton et al., 2018). NTFPs grown in gardens, orchards, swiddens and managed forest areas are biocultural keystones for socio-ecosystems managed to promote sustainable use (Ros-Tonen, 2000; Ros-Tonen & Wiersum, 2005; Shackleton et al., 2018). Rather than being vilified for causing deforestation, forest-dependent communities should have expanded tenure so that more forest is under socio-ecological management practices (Baragwanath & Bayi, 2020; Branford, 2020; FAO & FILAC, 2021).

NTFP extraction results in changes to forest ecosystems through selection for particular phenotypes, increased management of forest ecosystems and removal of individuals from the ecosystem (Arnold & Pérez, 2001; Shackleton et al., 2018). Extraction can be ecologically sustainable (Girão Rodrigues de Mello et al., 2020; Stanley et al., 2012). Management of forests for NTFPs can help to maintain ecosystems and promote biodiversity (FAO & FILAC, 2021; Shackleton et al., 2018). Sustainable management is associated with NTFP extraction for local use as part of diverse subsistence and income-generation activities and is guided by practices developed by forest-dependent communities. A study of NTFP sustainability found that in two-thirds of cases, extraction did not harm forest ecosystems (Stanley et al., 2012). However, this study did not integrate the socio-cultural contexts of NTFPs into the analysis (Girão Rodrigues de Mello et al., 2020), thereby replicating the assumptions of NTFP initiatives that identify NTFPs only as commodities that can be sold to economically incentivise the prevention of forest loss.

NTFP commercialisation poses risks of degradation to forest ecosystems. For example, açaí palms have been cultivated by Amazonian communities for thousands of years. They were traditionally planted in gardens or fallow areas with other fruit trees (Blaser et al., 2021). However, the recent harvest of açaí for commercial sale has driven extraction from unmanaged forests and has been found to cause significant changes to forest structure and species richness because lianas and understory plants are cleared to access açaí palm trees (Blaser et al., 2021).

When the demand and price of NTFPs increase, so does the risk of overexploitation—extracting a species faster than it can regenerate. Commercialisation can attract new extractors, including corporations, placing additional demands on the species. Overexploitation may also occur when prices fall and harvesters extract more to maintain their income (Girão Rodrigues de Mello et al., 2020). Overexploitation can result in decreased reproductive capacity and genetic stock of the NTFP species, population decline of the NTFP and other species that rely on it, removal of nutrients from the forest system and forest and biodiversity loss (Arnold & Pérez, 2001; Ros-Tonen, 2000; Shackleton et al., 2018). For example, demand for aguaje fruits led to a population decline of aguaje palms in Peru. Rather than climbing trees to harvest the fruits, harvesters cut down the trees to access the fruits because felling the tree enabled faster fruit collection (Penn, 2008). Felling the trees led to population decline and ultimately loss of access to the NTFP itself.

Species that are used for both timber and NTFPs are at high risk for overexploitation (Herrero-Jáuregui et al., 2013). Contrary to market theories that demand will lead extractors to optimise for the highest-value product, conflict generates competition between timber extractors and NTFP harvesters that leads to unsustainable harvest of the species for NTFPs and timber and may significantly damage the ecosystem (Herrero-Jáuregui et al., 2013; Rist et al., 2012). Larger actors may wrest control of an NTFP resource away from forest-dependent communities and then harvest both NTFPs and timber (Herrero-Jáuregui et al., 2013; Rist et al., 2012). Overexploitation and conflicts of use can be reduced through secure land tenure and protections for the use rights of forest-dependent communities, but robust policies are rare and not widely enforced (Baragwanath & Bayi, 2020; Girão Rodrigues de Mello et al., 2020; Herrero-Jáuregui et al., 2013). A common outcome is that NTFP commercialisation ends up marginalising the communities and harming the ecosystems that NTFP-based initiatives aim to help (Arnold & Pérez, 2001).

Ultimately, framing NTFPs only in economic terms excludes the socio-cultural roles of NTFPs and disregards local forest management practices (Girão Rodrigues de Mello et al., 2020). Forest-dweller management practices have been found to maintain or promote the growth of forest ecosystems (Branford, 2020; FAO & FILAC, 2021; Schuster et al., 2019), but these practices have been disregarded by forest managers, conservation NGOs and extractive industries. Communities have lost control of and access to the landscapes they managed and depended on for generations. Even conservation programmes that ask forest-dependent communities to apply and adapt their knowledge constrain community management practices to align with outside conservation and economic goals (Li, 2007; Whyte, 2016). Sustainable management of NTFP products and other forest resources that truly provides economic development to forest-dwelling communities will only be possible if conservation and development programmes integrate traditional and local knowledge.

3.3.3 The Bottom Line: Who Do Incentives Incentivise?

Perhaps the biggest challenge with using NTFP markets to incentivise conservation is that "incentives" beg the questions: "conservation of *what*?" and "conservation for *whom*?" NTFP-based initiatives assume that forest-dependent communities are responsible for deforestation and can therefore be incentivised to conserve, when, in reality, the majority of tropical forest deforestation is carried out by corporate entities while forest-dependent communities are disproportionately harmed by deforestation (Dove, 1994). Incentivisation schemes for conservation target the wrong actors. The people who harvest Brazil nuts are not the same people who cut down hundreds of acres of rainforest in the Amazon to raise cattle (Dove, 1994). As forest-dependent communities lack the political power and tenure rights to change land use, they cannot be expected to stop deforestation (Dove, 1994). Unless forest-dependent communities are permitted to control and develop their resources, NTFP trade will not succeed in incentivising conservation or bringing economic opportunity to forest communities (Baragwanath & Bayi, 2020; Branford, 2020).

Uninterrogated agendas and assumptions create conflicts and power disparities that undermine NTFP-based initiatives. Conservation NGOs identify NTFPs as economic resources that can be extracted without harming ecosystems, while development programmes see an opportunity for sustainable livelihoods. Government agencies view NTFP-based initiatives as a means to increase GDP and regulate forests. Timber corporations are interested in additional products to sell and the benefits of green marketing. Communities are concerned about maintaining access to territories, resources and cultural practices. The fact that forest-dependent communities' management practices may promote forest biodiversity does not signify that communities share the same goals for forest management as NGOs or government forest managers (Arnold & Pérez, 2001; Branford, 2020). As long as development is narrowly focused on improving economic conditions, programmes will continue to be blind to the external forces that degrade forests and impoverish forest communities (Dove, 1994; Girão Rodrigues de Mello et al., 2020; Rist et al., 2012).

3.4 Beyond Extraction: Other Approaches to NTFPs

The literature shows that NTFP-based initiatives fail to incentivise forest conservation or foster economic development and drive marginalisation of communities and overexploitation of ecosystems. This conclusion challenges the assumption that outsiders must do something *for* forest-dependent people and refocuses the discussion on what outsiders have done *to* forest-dependent people (Dove, 1994). "Forests are not degraded because forest peoples are impoverished; rather, forest peoples are impoverished by the degradation of their forests and other resources by external forces" (Dove, 1994). Reframing NTFPs as biocultural keystones (Shackleton et al., 2018) that provide insurance for forest-dependent peoples, enacting secure land tenure and resource rights and engaging local approaches to NTFP and forest management could provide avenues for forest-dependent communities to benefit from NTFP commercialisation while protecting ecosystems.

3.4.1 Local Management for Local Needs

NTFP extraction can meet local needs while not meeting conservation and development goals. For forest-dependent communities, NTFPs serve as "green social security" by filling subsistence gaps, provisioning materials like food or medicine that would need to be purchased and diversifying or supplementing income to cover needs like education or risks like changing agricultural production (Boon & Ahenkan, 2011; Global Forest Atlas, 2017; Pattanayak & Sills, 2001; Shackleton & Shackleton, 2004). NTFPs that are consistently available are more likely to be the basis for livelihoods, while less common or less available products are collected only seasonally or during lean times to fill subsistence and livelihood gaps (Belcher et al., 2005). Access to NTFPs for consumption and trade can particularly benefit groups who are excluded from mainstream economic activities, such as women, migrants and ethnic minorities (Shackleton & Shackleton, 2004).

Shackleton et al. (2018) propose understanding NTFPs as "biocultural keystone species". Biocultural keystone species play an outsized role in socio-ecosystems by providing services to people, for social and economic practices, and to other species (Shackleton et al., 2018). Palms are an important example of NTFP biocultural keystone species. Palms are used by people all over the world as sources of food, fibre and building materials and provide habitats and food sources for numerous animal species. Palms tend to be resilient to environmental shocks and are often available for people and animals in lean times when other resources are scarce (Shackleton et al., 2018). The açaí palm in the Amazon is an example. Its berries and stem "heart" can both be harvested for food, it grows easily, and it promotes the diversity of fruit-eating animals when planted in managed forest systems, such as gardens or fallow areas (Shackleton et al., 2018).

As deforestation and overexploitation drive degradation in forest-dependent communities' territories, cultivation and domestication of NTFP species are increasing to ensure continued access to NTFPs (Boon & Ahenkan, 2011; Hoch et al., 2012; Wood et al., 2016). A concern raised about NTFP cultivation is that domestication and agricultural intensification will lead to biodiversity loss. However, many NTFPs are already cultivated in gardens, agroforestry areas and swidden-fallow systems (Blaser et al., 2021). These systems can promote high agrobiodiversity in conjunction with intensification (Wood et al., 2016; Zimmerer, 2013). Similar to the case of açaí, a study in the Peruvian Amazon found that planting fallows with *umari* orchards rather than leaving the forest to regrow without management enabled harvesters to increase production and did not have any negative impacts on ecosystem services or biodiversity (Wood et al., 2016). This is an example of relying on local management practices—in this case, swidden cultivation—to grow NTFPs with an ecologically adapted and culturally appropriate method of intensification.

NTFPs are meeting the needs of forest-dependent communities, while conservation efforts fail to prevent forest loss and economic development programmes flounder. Forest-dependent communities should be provided with the tenure rights to develop their resources and be protected from the industrial drivers of deforestation (Blaser et al., 2021; Branford, 2020). NTFP-based initiatives can reframe NTFP extraction and cultivation as insurance for forest-dependent communities that promote locally led management that truly sustains people and ecosystems. Rather than contributing to marginalisation, NTFP-based initiatives can strengthen the position of forest-dependent communities by promoting systems like swidden-fallow cultivation that provide sustainable forest and NTFP management.

3.4.2 Integrated Management for Commercialisation

Integrated forest management (IFM) that focuses on NTFPs, rather than timber, can be an appropriate system to generate NTFPs for commercial sale while promoting conservation of biodiversity and ecosystem services, supporting livelihoods and following local management practices (Blaser et al., 2021; Herrero-Jáuregui et al., 2013; Hoch et al., 2012). IFM, which may also be called multiple-use management, requires extensive inventory by landowners—e.g. governments, corporations and communities—to understand the distribution of species present and how to manage species in a way that fosters symbiosis, biodiversity and production of timber and NTFPs (Blaser et al., 2021). IFM regimes should be designed based on specific cultural and ecological contexts, incorporating existing management approaches and strengthening land and use rights (Blaser et al., 2021). Importantly, IFM positions NTFPs one strategy to reduce deforestation (Blaser et al., 2021) in combination with other forest conservation efforts that engage local communities, and not as a standalone solution for conservation and sustainable development.

IFM can reduce conflicts of use by prioritising long-term sustainability and relying on local knowledge and management practices. For example, the most sustainable NTFPs are typically the fruits, flowers and seeds from perennial plants (Blaser et al., 2021). Forest can be managed to promote the growth and access to those NTFPs and selectively harvest timber from species that less readily produce NTFPs and are not biocultural keystone species. In the MAP region—Madre de Dios (Peru), Acre (Brazil) and Pando (Bolivia)—many communities depend on a variety of NTFPs, including Brazil nut, which is harvested in combination with timber (Rockwell et al., 2015). Sustainable management of Brazil nuts and timber is possible in the same region, as long as there is at least a 100 m buffer zone between the timber harvest area and the Brazil nut trees (Rockwell et al., 2015), which can be secured through IFM.

To avoid disenfranchising forest-dependent communities, people with local or indigenous knowledge of the ecosystem should lead in deciding which species should be used for particular activities and under what conditions (Klimas et al., 2012). The *Carapa guianensis* (andiroba tree) can be effectively harvested for both timber and seed oil if andiroba stands are divided into "specialized management units"—some for timber and some for seed oil (Klimas et al., 2012). Such a management system could disenfranchise local extractors by giving control of the NTFP resource to plantation managers. As an alternative strategy, local people with traditional knowledge were hired to identify the trees that were best for either NTFPs or timber (Klimas et al., 2012). This approach relied on local knowledge and returns authority to forest residents. Through IFM, forest managers can be trained in heterogeneous socio-ecological contexts and seek to complement, rather than override, local knowledge and management practices (Alexiades et al., 2013; Arnold & Pérez, 2001; Girão Rodrigues de Mello et al., 2020).

Market mechanisms that use NTFPs can be effective for forest conservation and sustainable management where rights are clear and forest-dependent communities are equal partners in designing projects. As payment for ecosystem services (PES) programmes have become increasingly popular, guidelines for ensuring appropriate consultation, benefit-sharing and partnership with forest-dependent communities have been developed (FAO & FILAC, 2021; FCPF & BioCF ISFL, 2020). Private capital channelled through PES can provide much-needed finance and support where government interventions are lacking. The robustness of PES varies widely, however. PES requirements must strive to integrate the leadership and knowledge of forest-dependent people. Under such conditions, NTFP commercialisation may more effectively provide income to forest-dependent communities and financial support for the management of ecosystems.

3.4.3 Regulation and Rights

In all cases—locally led forest management, swidden cultivation and agricultural intensification and commercial IFM—secure land tenure, clear rights and support of collective management must be in place (Baragwanath & Bayi, 2020; Blaser et al., 2021; Branford, 2020; Coomes, 2004; Dove, 1993; Girão Rodrigues de Mello et al., 2020). When governance conditions are not aligned with the socio-ecological context of forest products and users, long-term sustainability will not be achieved (Baragwanath & Bayi, 2020; Girão Rodrigues de Mello et al., 2020; Ros-Tonen, 2000; Shackleton et al., 2018). Forest-dependent communities require clear and secure rights to land, resources and knowledge to take the lead in developing management regimes and benefiting from the commercialisation of products (Branford, 2020). Government policy needs to robustly protect the rights of communities to develop and maintain control of valuable forest resources, through enforced land tenure, restrictions on industrial logging and monitoring of forest use (Blaser et al., 2021; Dove, 1993; FAO & FILAC, 2021). Conservation and development programmes that seek to integrate NTFPs can advocate for and strengthen the rights of local communities and commit to joint management led by existing practices. Ultimately, NTFP-based initiatives may be used by forest-dependent communities to assert more control over territories that are threatened by other forms of resource extraction, like deforestation, and to create more political power for themselves and the communities they represent.

3.5 Conclusion

Tracing the literature on NTFPs finds that NTFP commercialisation as an incentive for forest conservation is not "benign", as initiatives propose, but has marginalised forest-dependent people and led to overexploitation of forests. NTFP-based initiatives assume that forest-dependent people must be incentivised to halt deforestation, disregard existing local practices and place blame for deforestation on those people who are both most impacted by it and possess the knowledge of how to manage forests sustainably. NTFPs are commercially and culturally important products. Forest-dependent communities should be able to profit from the extraction and sale of those products—as well as choose to develop, use and consume NTFPs for other purposes.

A prerequisite to any NTFP-based programme should be to understand existing social, cultural and economic systems that use and rely on NTFPs (Blaser et al., 2021; Dove, 1994; Girão Rodrigues de Mello et al., 2020; Shackleton et al., 2018). Non-timber forest products are local, contextual and culturally specific objects whose value is derived from uses forest dweller communities discovered and developed. This definition of NTFPs makes it clear that NTFP resources should be managed by local knowledge holders with a recognition of the economic, subsistence and cultural roles of NTFPs in the lives of forest-dependent peoples.

Acknowledgements Thank you to Dr. Norfaryanti Kamaruddin for inviting me to participate in this project and to the editors for your critiques that strengthened this chapter. Thanks also to Dr. Michael Dove for providing the course where I developed this research and for feedback on early drafts.

References

Aguirre Mendoza, Z., & Aguirre Mendoza, L. (2021). Estado actual e importancia de los productos forestales no maderables. *Bosques Latitud Cero, 11*, 71–82.

Ahenkan, A., & Boon, E. (2010). Commercialization of non-timber forest products in Ghana: Processing, packaging and marketing. *Journal of Food, Agriculture and Environment, 8*, 962–969.

Alexiades, M. N., Peters, C. M., Laird, S. A., et al. (2013). The missing skill set in community management of tropical forests. *Conservation Biology, 27*, 635–637. https://doi.org/10.1111/cobi.12040

Ambrose-Oji, B. (2003). The contribution of NTFPs to the livelihoods of the 'forest poor': Evidence from the tropical forest zone of south-west Cameroon. *International Forestry Review, 5*, 106–117.

Arnold, D. (2000). "Illusory riches": Representations of the tropical world, 1840–1950. *Singapore Journal of Tropical Geography, 21*, 6–18.

Arnold, J. E. M., & Pérez, M. R. (2001). Can non-timber forest products match tropical forest conservation and development objectives? *Ecological Economics, 39*, 437–447. https://doi.org/10.1016/S0921-8009(01)00236-1

Baragwanath, K., & Bayi, E. (2020). Collective property rights reduce deforestation in the Brazilian Amazon. *Proceedings of the National Academy of Sciences of the United States of America, 117*, 20495–20502. https://doi.org/10.1073/pnas.1917874117

Belcher, B. M. (2003). What isn't an NTFP? *International Forestry Review, 5*, 161–168.

Belcher, B. M., Ruíz-Pérez, M., & Achdiawan, R. (2005). Global patterns and trends in the use and management of commercial NTFPs: Implications for livelihoods and conservation. *World Development, 33*, 1435–1452. https://doi.org/10.1016/j.worlddev.2004.10.007

Blaser, J., Frizzo, J. C., & Norgrove, L. (2021). *Not only timber: The potential for managing non-timber forest products in tropical production forests—A comprehensive literature review*. International Tropical Timber Organization (ITTO) and Precious Forests Foundation.

Boon, E., & Ahenkan, A. (2011). NTFPs: Clearing the confusion in semantics. *Journal of Human Ecology, 33*, 1–9.

Branford, S. (2020). Indigenous best Amazon stewards, but only when property rights assured: Study. In *Mongabay Environmental News*. Retrieved May 5, 2021, from https://news.mongabay.com/2020/08/indigenous-best-amazon-stewards-but-only-when-property-rights-assured-study/

Byg, A., & Balslev, H. (2006). Palms in indigenous and settler communities in Southeastern Ecuador: Farmers' perceptions and cultivation practices. *Agroforestry Systems, 67*, 147–158. https://doi.org/10.1007/s10457-005-1704-1

Coomes, O. T. (2004). Rain forest "conservation-through-use"? Chambira palm fibre extraction and handicraft production in a land-constrained community, Peruvian Amazon. *Biodiversity & Conservation, 13*, 351–360. https://doi.org/10.1023/B:BIOC.0000006503.90980.e8

Coomes, O. T., Takasaki, Y., & Rhemtulla, J. M. (2016). Forests as landscapes of social inequality: Tropical forest cover and land distribution among shifting cultivators. *Ecology and Society, 21*, 15.

de Beer, J. H., & McDermott, M. J. (1996). *The economic value of non-timber forest products in Southeast Asia*. IUCN.

Dove, M. R. (1993). A revisionist view of tropical deforestation and development. *Environmental Conservation, 20*, 17–24. https://doi.org/10.1017/S0376892900037188

Dove, M. R. (1994). *Marketing the rainforest: 'Green' panacea or red herring?* AsiaPacific Issues – Analysis from the East-West Center, No. 13. http://scholarspace.manoa.hawaii.edu/handle/10125/3777, https://scholarspace.manoa.hawaii.edu/server/api/core/bitstreams/af0ab573-18b4-4aad-8149-21085ef696dd/content

Dove, M. R. (2011). *The banana tree at the gate*. Yale University Press.

Dove, M. R., & Kammen, D. (1997). The epistemology of sustainable resource use: Managing forest products, swiddens, and high-yielding variety crop. *Human Organization, 56*, 91–101.

Drahos, P., & Frankel, S. (2012). *Indigenous peoples' innovation: Intellectual property pathways to development*. Australian National University E Press.

FAO & FILAC. (2021). *Forest governance by indigenous and tribal peoples. An opportunity for climate action in Latin America and the Caribbean*. FAO.

FCPF & BioCF ISFL. (2020). *Designing benefit sharing arrangements: A resource for countries*. Retrieved August 9, 2021, from https://www.forestcarbonpartnership.org/bio-carbon/en/index.html

Girão Rodrigues de Mello, N., Gulinck, H., Van den Broeck, P., & Parra, C. (2020). Social-ecological sustainability of non-timber forest products: A review and theoretical considerations for future research. *Forest Policy and Economics, 112*, 102109. https://doi.org/10.1016/j.forpol.2020.102109

Global Forest Atlas. (2017). *Non-timber forest products, rubber, and medicines*. Yale School of Forestry and Environmental Studies. Retrieved November 11, 2017, from https://globalforestatlas.yale.edu/amazon/forests-and-logging/non-timber-forest-products

Herrero-Jáuregui, C., Guariguata, M. R., Cárdenas, D., et al. (2013). Assessing the extent of "conflict of use" in multipurpose tropical forest trees: A regional view. *Journal of Environmental Management, 130*, 40–47. https://doi.org/10.1016/j.jenvman.2013.08.044

Hoch, L., Pokorny, B., & Jong, W. (2012). Financial attractiveness of smallholder tree plantations in the Amazon: Bridging external expectations and local realities. *Agroforestry Systems, 84*, 361–375. https://doi.org/10.1007/s10457-012-9480-1

Klimas, C. A., Kainer, K. A., & de Oliveira Wadt, L. H. (2012). The economic value of sustainable seed and timber harvests of multi-use species: An example using *Carapa guianensis*. *Forest Ecology and Management, 268*, 81–91. https://doi.org/10.1016/j.foreco.2011.03.006

Li, T. M. (2007). *The will to improve: Governmentality, development, and the practice of politics*. Duke University Press Books.

Li, T. M. (2014). *Land's end: Capitalist relations on an indigenous frontier*. Duke University Press.

Osborne, T. (2015). Tradeoffs in carbon commodification: A political ecology of common property forest governance. *Geoforum, 67*, 64–77. https://doi.org/10.1016/j.geoforum.2015.10.007

Padoch, C. (1988). Chapter 6. The economic importance and marketing of forest and fallow products in the Iquitos region. In *Swidden-fallow agroforestry in the Peruvian Amazon* (pp. 75–89). The New York Botanical Garden.

Pattanayak, S. K., & Sills, E. O. (2001). Do tropical forests provide natural insurance? The microeconomics of non-timber forest product collection in the Brazilian Amazon. *Land Economics, 77*, 595–612. https://doi.org/10.2307/3146943

Penn, J. W. (2008). Non-timber forest products in Peruvian Amazonia: Changing patterns of economic exploitation. *Focus on Geography, 51*, 18–25. https://doi.org/10.1111/j.1949-8535.2008.tb00222.x

Peters, C. M., Gentry, A. H., & Mendelsohn, R. O. (1989). Valuation of an Amazonian rainforest. *Nature, 339*, 655–656.

Quaedvlieg, J., Roca, M. G., & Ros-Tonen, M. A. F. (2014). Is Amazon nut certification a solution for increased smallholder empowerment in Peruvian Amazonia? *Journal of Rural Studies, 33*, 41–55. https://doi.org/10.1016/j.jrurstud.2013.10.004

Rist, L., Shanley, P., Sunderland, T., et al. (2012). The impacts of selective logging on non-timber forest products of livelihood importance. *Forest Ecology and Management, 268*, 57–69. https://doi.org/10.1016/j.foreco.2011.04.037

Rockwell, C. A., Guariguata, M. R., Menton, M., Quispe, E. A., Quaedvlieg, J., Warren-Thomas, E., Silva, H. F., Rojas, E. E. J., Arrunátegui, J. A. H. K., Vega, L. A. M., Vera, O. R., Hancco, R. Q., Tito, J. F. V., Panduro, B. T. V., & Salas, J. J. (2015). Nut Production in Bertholletia excelsa across a logged forest mosaic: Implications for multiple forest. *PLoS One, 10*(8), e0135464. San Francisco. https://doi.org/10.1371/journal.pone.0135464. https://www.proquest.com/docview/1705068568/abstract/B411D3849E224AC9PQ/1

Ros-Tonen, M. A. F. (2000). The role of non-timber forest products in sustainable tropical forest management. *Holz als Roh- und Werkstoff, 58*, 196–201. https://doi.org/10.1007/s001070050413

Ros-Tonen, M. A. F., & Wiersum, K. F. (2005). The scope for improving rural livelihoods through non-timber forest products: An evolving research agenda. *Forests, Trees and Livelihoods, 15*, 129–148. https://doi.org/10.1080/14728028.2005.9752516

Schuster, R., Germain, R. R., Bennett, J. R., et al. (2019). Vertebrate biodiversity on indigenous-managed lands in Australia, Brazil, and Canada equals that in protected areas. *Environmental Science & Policy, 101*, 1–6. https://doi.org/10.1016/j.envsci.2019.07.002

Shackleton, C., & Shackleton, S. (2004). The importance of non-timber forest products in rural livelihood security and as safety nets: A review of evidence from South Africa. *South African Journal of Science, 100*, 658–664.

Shackleton, C., Ticktin, T., & Cunningham, A. (2018). Nontimber forest products as ecological and biocultural keystone species. *Ecology and Society, 23*, 230422. https://doi.org/10.5751/ES-10469-230422

Shanley, P., Luz, L., & Swingland, I. R. (2002). The faint promise of a distant market: A survey of Belém's trade in non-timber forest products. *Biodiversity & Conservation, 11*, 615–636. https://doi.org/10.1023/A:1015556508925

Stanley, D., Voeks, R., & Short, L. (2012). Is non-timber forest product harvest sustainable in the less developed world? A systematic review of the recent economic and ecological literature. *Ethnobiology and Conservation, 1*, 1–39.

Whyte, K. P. (2016). What do indigenous knowledges do for indigenous peoples? In M. K. Nelson & D. Shilling (Eds.), *Keepers of the Green World: Traditional ecological knowledge and sustainability* (pp. 57–82). Cambridge University Press.

Wood, S. L. R., Rhemtulla, J. M., & Coomes, O. T. (2016). Intensification of tropical fallow-based agriculture: Trading-off ecosystem services for economic gain in shifting cultivation landscapes? *Agriculture, Ecosystems & Environment, 215*, 47–56. https://doi.org/10.1016/j.agee.2015.09.005

Woods, K. M., & Naimark, J. (2020). Conservation as counterinsurgency: A case of ceasefire in a rebel forest in southeast Myanmar. *Political Geography, 83*, 102251. https://doi.org/10.1016/j.polgeo.2020.102251

Zimmerer, K. S. (2013). The compatibility of agricultural intensification in a global hotspot of smallholder agrobiodiversity (Bolivia). *Proceedings of the National Academy of Sciences of the United States of America, 110*, 2769–2774.

Characterizing and Assessing Forest Density and Productivity of Ulu Muda Forest Reserve Based on Satellite Imageries

4

Sheriza Mohd Razali, Nor Rohaizah Jamil, Mohd Sofiyan Sulaiman, and Maizatul Azwa Radzi

Abstract

The NPP was calculated for 2-m resolution image of Ulu Muda Forest Reserve (UMFR), in Sik, North of Peninsular Malaysia. The GeoEye-1 image was first preprocessed, and land use was classified based on object-based image analysis (OBIA) based on PCI Geomatics Catalyst Professional image processing software. Attributes were segmented by employing three segmentation methods, namely, finer, 50 scale; moderate, 200; and coarse, 350. Based on accuracy assessment, a moderate scale of 200 showed the best kappa with 0.67, whereas for finer and coarse were 0.44 and 0.27, respectively. The moderate segmentation method showed a moderate number of attributes that sufficiently assist in collecting accuracy sampling that resulted in a higher kappa coefficient in the study. Biophysical indices, such as Absorbed Photosynthetically Active Radiation (APAR), Normalized Difference Vegetation Index (NDVI) and the fraction of Photosynthetically Active Radiation (fAPAR), were calculated for the study based on the satellite images. The study showed that the coarse method of NDVI had the highest mean value of 0.709, followed by 0.698 for moderate and 0.966 for finer method. A high NDVI value indicated that the area in UMFR is covered by high-density vegetation dominated by lowland forest. Meanwhile, the

S. M. Razali (✉) · M. A. Radzi
Institute of Tropical Forestry and Forest Products (INTROP), Universiti Putra Malaysia (UPM), Serdang, Selangor, Malaysia
e-mail: sheriza@upm.edu.my

N. R. Jamil
Department of Environment, Faculty of Forestry and Environment, Universiti Putra Malaysia (UPM), Serdang, Selangor, Malaysia

M. S. Sulaiman
Faculty of Ocean Engineering Technology and Informatics, Universiti Malaysia Terengganu, Kuala Nerus, Terengganu, Malaysia

Z. Samdin et al. (eds.), *Tropical Forest Ecosystem Services in Improving Livelihoods For Local Communities*, https://doi.org/10.1007/978-981-19-3342-4_4

recorded NPP ranged between 6.7 g C m^{-2} $month^{-1}$ and 300.04 g C m^{-2} $month^{-1}$, with a mean value of 231.85 g C m^{-2} $month^{-1}$ for the study area. Satellite remote sensing allows for the estimation of NPP while also generating land use/land cover and forest density estimates for lowland forests in the peninsular. The findings indicate the presence of intricacies between NDVI, forest density and land cover in explaining NPP variations within this type of forest. The Ulu Muda FR community assessed by the study extracted no major NPP. This type of research can be used in forest management planning in the forestry department to improve forest extraction policy.

Keywords

Remote sensing · NPP · NDVI · Ulu Muda Forest Reserve · Productivity

4.1 Net Primary Productivity

Primary production is the rate of solar energy converted to plant biomass through photosynthesis (Indiarto & Sulistyawati, 2013). Gross Primary Productivity (GPP) is important to measure because it indicates the tree's capability to produce biomass at a specific time; however, understanding the tree's vigour with the given monthly temperature and solar radiation is not possible, certainly not in a short period of time. The GPP calculates the total amount of solar energy converted to biomass (Indiarto & Sulistyawati, 2013). Net primary productivity (NPP), on the other hand, is the net flux of carbon from the atmosphere into green plants per unit of time (Zhang et al., 2017). It is also known as an important index for evaluating the carbon cycling in forest ecosystems (Muhamad, 2010).

The GPP estimation is practical when afforded with a longer period of time, especially when combined with a temporal approach offered by satellite remote sensing. The introduction of MODIS NPP and GPP products in recent years has made the phenological estimation of global terrestrial possible. When employed with the specific model of Biome-BGC carbon cycle, the products showed a promising outcome (Turner et al., 2006). To date, quantification of GPP has been conducted using satellite remote sensing of Sentinel-2 images for a short rotation of forest plantations in Belgium (Maleki et al., 2020).

A comparative study on NPP changes by Shao and Zeng (2017) puts global tropical forests under the spotlight. After that, further studies on tropical forest became sparse, with more studies on NPP for plantation emerging instead of studies on natural forest land. Therefore, this study aims to assess NPP for UMFR. To do this, the study employed an NPP equation using satellite remote sensing of GeoEye-1 data of Sungai Teliang, UMFR, in Sik, Kedah, Malaysian Peninsula. The forest is one of the tropical evergreen forests in Southeast Asia with a high biodiversity value. In this study, forest density mapping was performed to obtain land cover information for the study area, which was then correlated with the NPP and NDVI values of the study. Three segmentation methods were designated in developing the land cover maps, namely, finer, medium and coarse. The finer scale is used as reference and validation while the other two scales, i.e. moderate and coarse, as a test method to derive the forest attributes.

NPP has been proven to be very useful carbon cycling assessment of forest ecosystems as demonstrated in a study in rocky areas in southwestern China, which lie in the transition zone of the Qinghai–Tibet Plateau and the Yangtze River valley in the middle and lower reaches (Zhang et al., 2017). Other study showed that NPP is varied between the southern, central and northern sites found in a study in boreal forest, where different climate-driven processes regulate forest growth and C uptake. At the same time, the study found water availability appears to be the critical factor limiting forest NPP on southern site in that tested site (Peng & Apps, 1999). NPP estimation permits knowledge about rainfall amount, maximum temperature, solar radiation and potential evaporation.

4.2 Normalized Difference Vegetation Index (NDVI)

NDVI is an essential indication for estimating productivity, whether it is derived at a higher resolution from image or using products such as MODIS Net Primary Production Yearly L4 Global 1 km (MYD17A3). NDVI is a vegetation index that is used to understand drought and soil nutrient loss and, in general, to identify vegetation stress in the vegetation (Luus & Kelly, 2008). The NPP at the regional level varies depending on the temperature. For instance, the vegetation productivity for a region in a cold area, such as the Pan-Arctic, is lower compared to temperate forests, due to the low temperature that limits plant metabolic activity (Kimball et al., 2006). Many countries are currently estimating NPP for their forests to evaluate carbon cycling in their forest ecosystems. Muhamad (2010), for example, estimated NPP for Kalimantan using MODIS products. Generally, MODIS products can be used to estimate GPP and NPP for various types of forests, namely, evergreen needleleaf forest (EN), evergreen broadleaf forest (EB), deciduous needleleaf forest (DN), deciduous broadleaf forest (DB) and mixed forest (Hashimoto et al., 2012).

Meanwhile numerous researches have demonstrated the applicability of the Normalized Difference Vegetation Index (NDVI) in estimating GPP phenology in short-rotation plantation. For example, the GPP in Changbai Mountain, China, was estimated using MODIS products of land cover and Enhanced Vegetation Index (EVI)/NDVI products (Zhang et al., 2019). In terms of image resolution, it was discovered that using recent and high-resolution images for NPP estimation provided greater benefits to local authorities than using a larger scale.

4.3 Motivation of the Study

As a result, this study estimates NPP for the Ulu Muda FR using OBIA image analysis of remote sensing imagery while also estimating forest density. In addition, the study determined the final NPP value and land use classes for Ulu Muda

FR. Finally, the study connected the NPP to the socio-demographics of the community in districts near Sik, specifically Baling, Kulim and Kuala Muda. The study assumed that the process is part of their socio-economic process that at the same time has interaction on NPP spatial area. It is critical to investigate the relationship between forest density and NPP and to assess its effects on the study area community. In other studies, Razali et al. (2015) assessed the human impact on NPP in the tropical forest in Malaysian Peninsula, while Vu et al. (2014) estimated NPP as a proxy for persistent decline or improvement to reflect previous land degradation in Vietnam.

4.4 Methodology

4.4.1 Study Site

The study was conducted in Sungai Teliang River in UMFR, Sik, Kedah, Malaysian Peninsula. The UMFR is part of the Greater Ulu Muda Forest Complex (UMFC) in north-eastern Kedah (Aik et al., 2017). It consists of about 105,060 ha of forest, which lays on the lowland to hill dipterocarp forest and riverine forest (Sharma et al., 2005). The forest reserve is a protected forest serving as a reservoir for Muda, Pedu and Ah Ning Lake (Fig. 4.1). The forest has a unique grassy environment of Sira Bongor and Sira Keladi that are dynamic habitat patches created and maintained by elephants and other larger herbivores (Chew et al., 2014). Physical observation from the study found that a small herd of elephants was heard walking, feeding, drinking, bathing and actively vocalizing from 22:00 to 02:00 h at Sira Air Hangat on 26—27 February 2014. They were moving in the forest surrounding the salt lick and spent about an hour splashing around the lower stretches of the stream before moving on towards Sungai Muda (Chew et al., 2014). This makes the forest a rich biodiversity habitat for wildlife and a critical water catchment for Kedah, Perak and Penang Island. Meanwhile, the UMFR forms part of a massive water catchment in Kedah, in the north-western part of Peninsular Malaysia, that drains either into the Muda dam or into Sungai Muda downstream from the dam (Miniandi et al., 2021). Annual precipitation recorded for the study area ranged from 1972.2 to 2478.0 mm from 1989 to 1997. The monthly mean averaged 177.43–233.43 mm, indicating an increasing precipitation pattern in Ulu Muda FR. Meanwhile, monthly precipitation ranged from 164.5 to 240.4 mm between 1989 and 2018 (Fig. 4.2).

4.4.2 NPP and Social Demographic Ulu Muda FR Community

Indirect social demographic profile of the Ulu Muda FR community analysis was retrieved and referred from interviewed session conducted by Mei (2020) study that was published in 2020. The study evaluated socio-economic and livelihood status of the inland fisherman around Muda River basin that connected to the forest. A total 46 fisherman respondents were selected randomly from the list provided by the

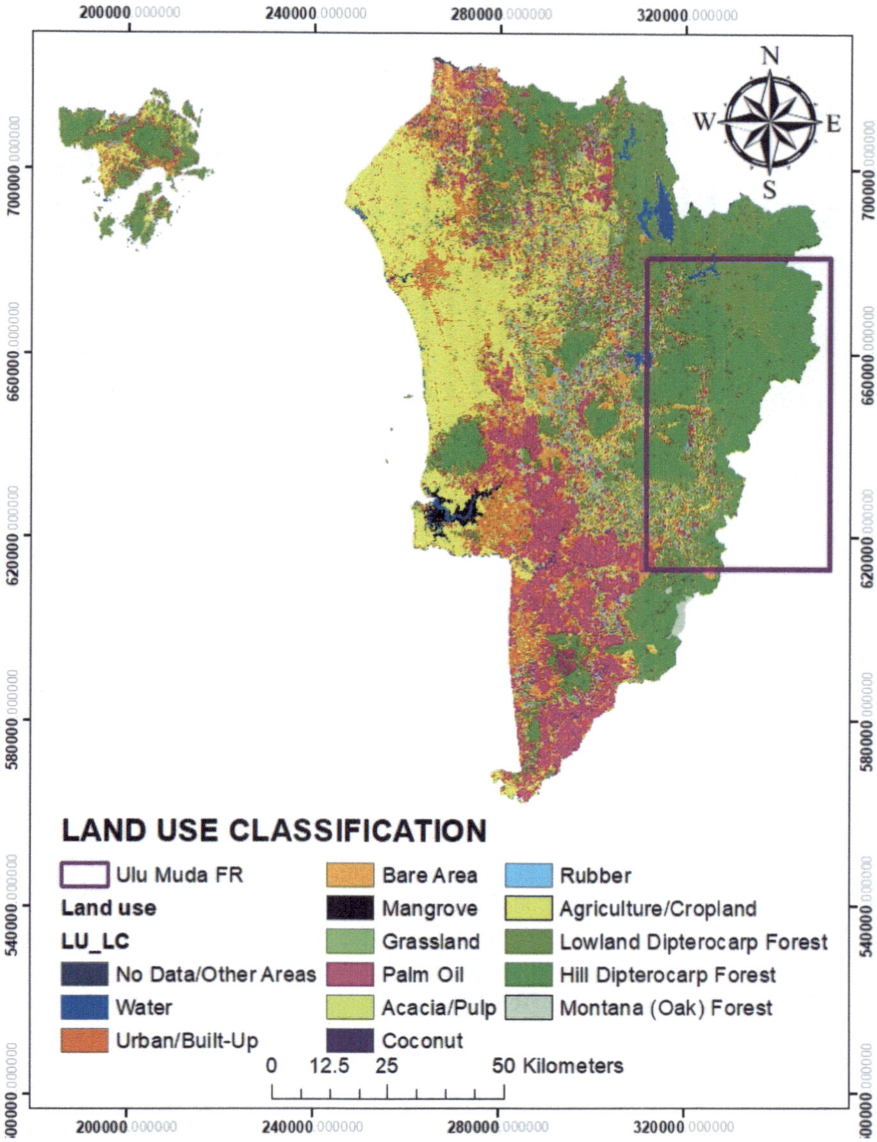

Fig. 4.1 Location of the study area, Ulu Muda Forest Reserve within Kedah map

Department of Fisheries. The article started the data collection in November 2018. The fisherman demographic information was collected and was analysed using Statistical Package for Social Science (SPSS). About 46 fishermen were successfully interviewed from November 2018 until March 2019. Areas that were included in the interviewed were listed below. In general, this interview covered an upstream area which was Kampung Kuala Lesung (Baling district) and mid-stream area of Muda

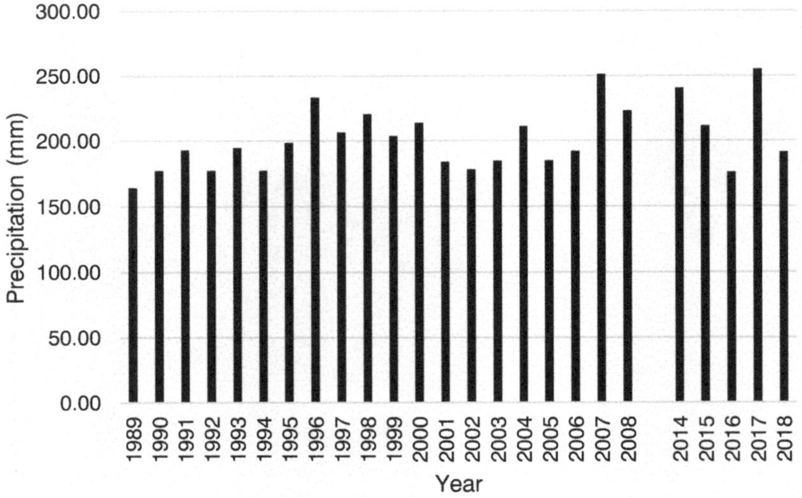

Fig. 4.2 Monthly precipitation for UMFR from 1989 to 2018

River basin that was Muda dam, Gubir (Sik district) and Kampung Ujung Padang (Kulim district). Finally, the study was also included downstream area that was Kampung Titi Merdeka (Kuala Muda district) (please refer Kedah District map in Fig. 4.2).

4.4.3 Satellite Data

The GeoEye-1 data for Sungai Teliang was acquired from image data at 0.46-m panchromatic (black and white), and 1.84-m multispectral bands were used for the NPP estimation, dated 5 April 2018 (Fig. 4.3). The ArcGIS pixel size was approximately 2 m. The GeoEye-1 satellite sensor features the most sophisticated technology ever used in a commercial remote sensing system. It is optimized for large projects, as it can produce over 350,000 km^2 of pan-sharpened multispectral satellite imagery every day. GeoEye-1 has been flying at an altitude of about 681 km and can produce imagery with a ground sampling distance of 46 cm, meaning it can detect objects of that diameter or greater (Satellite Imaging Corporation, 2017). The satellite image employed in this study consists of four bands, as shown in Table 4.1.

4.4.4 Data Analysis

The study applied ArcGIS version 10.8.2 and PCI Geomatics Catalyst Professional software. The flow data analysis from preprocessing, classification and land use mapping is shown in Fig. 4.4.

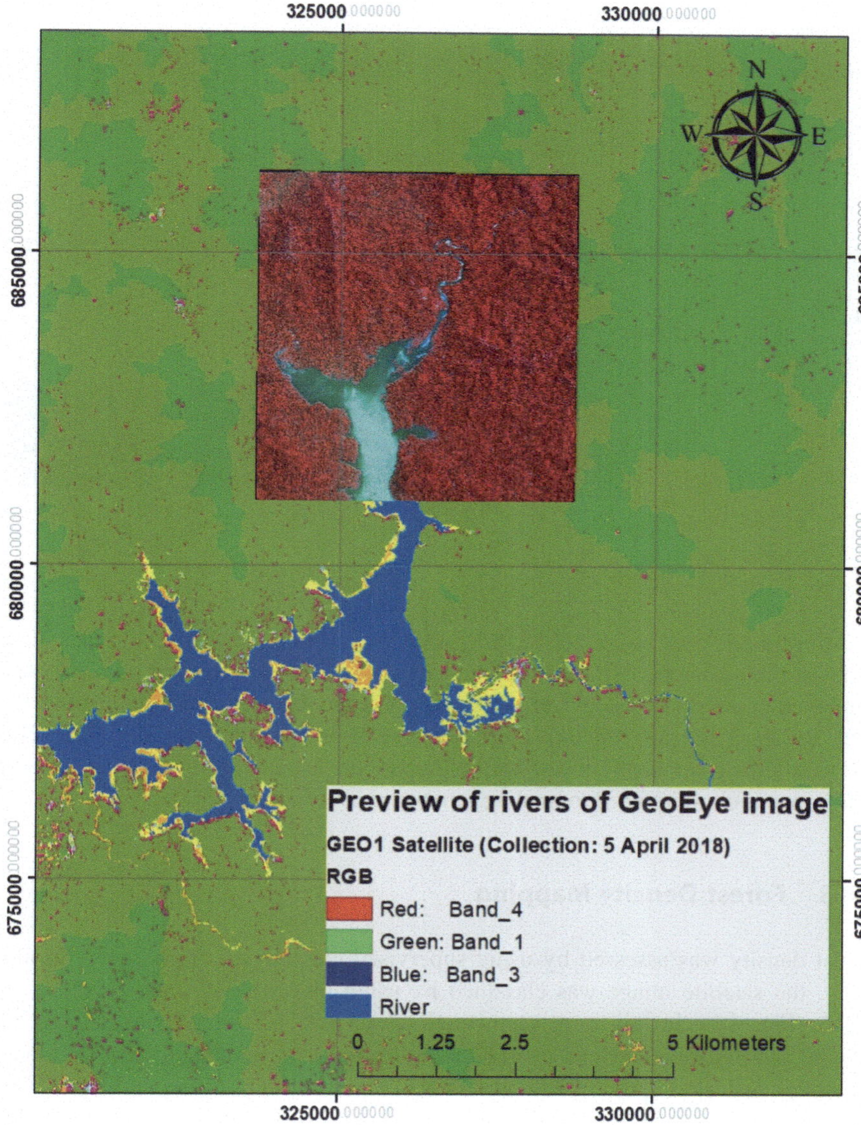

Fig. 4.3 GeoEye-1 preview of Muda Dam and Teliang River dated 5 April 2018

Table 4.1 Spectral range of GeoEye-1 satellite image

Bands	Spectral range (nm)
Blue	450–510
Green	510–580
Red	655–690
Near infrared	780–920

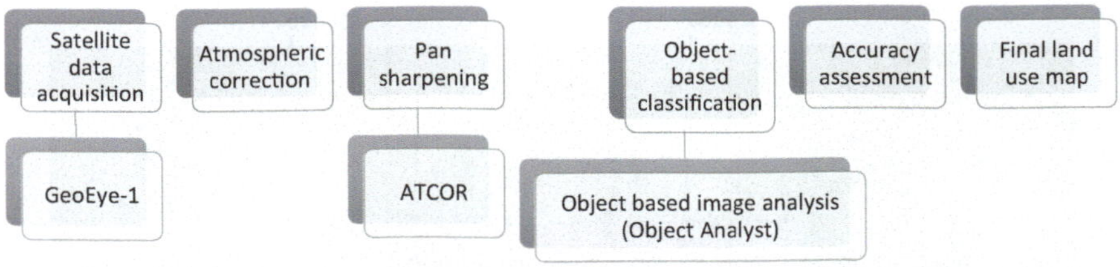

Fig. 4.4 Steps in land use mapping from preprocessing, classification to the final land use map

Fig. 4.5 Image preprocessing using ATCOR of PCI Geomatics Catalyst Professional software

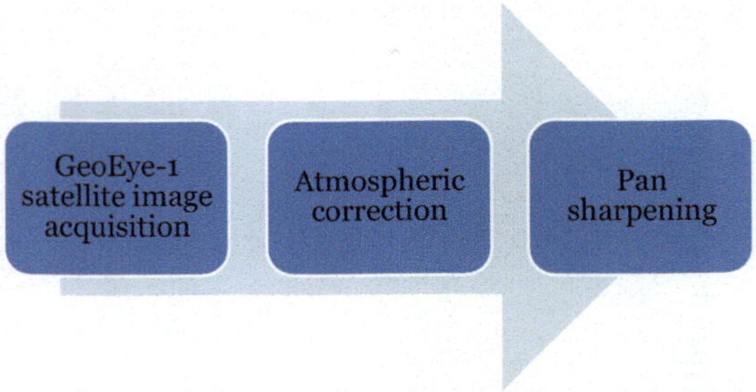

Atmospheric correction tool in PCI Geomatica, which allows users to execute a variety of atmospheric corrections in the simplest and fastest method possible, was employed in the study. The wizard fills in most of the required parameters automatically using image information and walks the user through each key step. The software's focus application was used to prepare the data, and then ATCOR ground reflectance tools were used to analyse atmospheric correction (Fig. 4.5).

4.4.5 Forest Density Mapping

Forest density was assessed by using supervised classification techniques. In this study, the satellite image was classified by using object-based image analysis or OBIA (Fig. 4.6). In this step, segmentation was first performed with a scale of 50, 200 and 350 characterized by a shape of 0.5 and compactness of 0.5. After that, the study calculated the statistical attributes of the segmentation image. Meanwhile, the scale can be modified to fit the land cover types, such as a finer scale for a city with buildings. However, in forest areas with few land cover categories, a greater scale is required for segmentation in order to construct a larger polygon for all the features. The training pixels data were then classified using support vector machine (SVM) classifier; then, accuracy assessment was conducted for all the training and classified images. Accuracy assessment was conducted by assessing overall accuracy and producer's and user's accuracy of the final land use maps based on the three segmentation scales. The maps' confusion matrices and Cohen's Kappa coefficient (Cohen, 1960) were calculated from the error images for each of the three methods

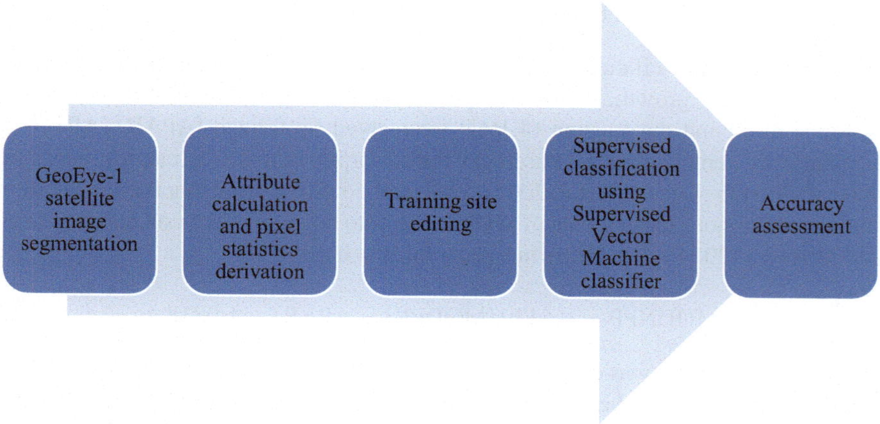

Fig. 4.6 Forest density mapping using OBIA and support vector machine classification

and then reported in an error matrix. After that, the classified polygon was converted into points to calculate the GPP for all the land cover points using ArcGIS software.

4.4.5.1 Forest Density Distribution by Area

Layer properties in ArcGIS software is a tool that uses layer symbology to present quantities of the polygons in the feature layer. Values for fields were chosen, and the colour ramp with suitable range colour was selected. The number of classes chosen was five as the default class due to the medium range of attributes present in the data for all segment scales. Forest density distribution by area was applied to all the segmentation techniques. The study applied graduated colours with Natural Breaks, and five classes were chosen. The chosen colour ramps were red, green and yellowish colour. They are then imported onto all the segmentation for further analysis. The objective of the analysis was to obtain polygon ranges for all the segmentation.

Meanwhile, for further analysis, attributes for more than 1 ha area were extracted by applying select by attributes and added to (area ha > 1) in attributes dialog. The objective of this analysis was to obtain polygon within ranges of more than 1 ha.

4.4.5.2 NPP Equation

Biophysical indices were estimated by using satellite image of GeoEye-1. The indices were included in calculating NPP for the study area. Generally, the NPP can be estimated by reducing the value to 50% of the GPP value. This equation has been employed by many researchers in estimating NPP for tropical forests (Faidi et al., 2010; Girardin et al., 2010; Rasib et al., 2008; Razali et al., 2015). Therefore, in this study, GPP was first quantified using an equation from Xiao et al. (2004). Referring to the study, the GPP is calculated as the following:

$$\text{GPP} = \text{fAPAR} \times \text{PAR} = \varepsilon \times \text{APAR} \tag{4.1}$$

where ε is light use efficiency (ε), which is an important element of the radiation regime for tree growth, incoming photosynthetically active radiation (PAR) ($g\ MJ^{-1}\ m^{-2}$) and the fraction of PAR intercepted by foliage fAPAR. fAPAR is estimated by remote sensing based on the absorption of a tree canopy between 400 and 700 nm (Luus & Kelly, 2008). This method differs from that used in another study using remote sensing data of MODIS satellite image products such as Pachavo and Murwira (2014). The following equation was used:

$$\text{MODNPP} = 0.5139\,(\text{MODPRI} \times \text{APAR}) - 1.9818 \tag{4.2}$$

In this study, PAR was estimated from 50% of incoming solar radiation based on the data collected from the meteorological station in the forest reserve. Solar radiation was collected from 1979 until 2014, which is expressed as $W\ m^{-2}$. The study derived the value for absorbed fraction of photosynthetically active radiation (APAR) ($g\ MJ^{-1}$) by multiplying the two most important elements of radiation, fAPAR and PAR (Coops et al., 2010).

Solar radiation in Malaysia is estimated at around 400–600 $MJ\ m^{-2}$, which is higher during the monsoon season of the north-east when the wind direction comes from Central Asia to the South China Sea. After that, the wind goes to Australia between November and March at the final stage (Abdullah et al., 2019). It indicates that Malaysia has high photosynthetic activity in the forests. Daily solar radiation in the forest reserve averaged 18.36 $MJ\ m^{-2}\ day^{-1}$ between 1978 and 2014. Solar radiation is required for the growth of vegetation on the earth, and many industries used photovoltaic (PV) to convert it into energy for usage. Finally, to calculate the LUE, we referred to a study by Handcock and Csillag (2004):

$$\text{LUE} = 0.8932 + T_{\text{month}} + 0.0015\,(\text{PRECIP}_{\text{month}}) - 0.002\,(\text{GDD}) \tag{4.3}$$

where the LUE ($g\ MJ^{-1}$) equation was referenced in Band et al. (1999) in which the study had developed based on the data obtained from several sources at the Meteorological Department for the Ulu Muda dam. $\text{PRECIP}_{\text{month}}$ is daily precipitation from 1989 until 2014 that was averaged into monthly precipitation in millimetre (mm) collected from the Ulu Muda weather station derived from the Malaysia Meteorological Department (2018). Meanwhile, the value for T_{month} is the monthly temperature derived for Ulu Muda station, expressed in maximum and minimum temperature for daily readings. Minimum and maximum values can show the variation of temperature over a single day. The data selected for the equation is the maximum value recorded in Celsius (°C) calculated from 1979 data until 2014. The air temperature during the north-east monsoon season in Malaysia tends to lower with solar radiation to below 5000 $W\ m^{-2}\ day^{-1}$ (Muzathik et al., 2010).

The GDD is the average of growing degree days calculated by finding the difference between the mean and minimum temperature values (Sunaryathy et al., 2012), which is also used to evaluate the growth response of the trees. For this study, the GDD was 9.48 °C. The fAPAR was based on a study by Goward et al. (1994). In this study, the value for fAPAR was derived as:

$$\text{fAPAR} = 1.2 \times \text{NDVI} - 0.04 \tag{4.4}$$

The NDVI was derived by using two bands in the satellite image as shown below, as referenced to a study by Rouse et al. (1973):

$$\text{NDVI} = \frac{\rho_{\text{NIR}} - \rho_{\text{Red}}}{\rho_{\text{NIR}} + \rho_{\text{Red}}} \tag{4.5}$$

where ρ_{NIR} is the reflectance of the WorldView image at 0.77–0.895 nm (near-infrared band, referred to as band 4) and ρ_{Red} is the reflectance of the satellite image at 0.63–0.690 nm (red band referred to as band 3). The NDVI value ranges from -1 and 1, where -1 represents bare soil and 1 as dense vegetation (Senna, 2005). Using ArcGIS software, Eqs. (4.1)–(4.4) and Eq. (4.6) were employed to calculate the NPP.

After that, the NDVI was calculated using Spatial Analyst map algebra in ArcGIS using the following equation:

$$\text{NDVI ArcGIS} = \text{Float}\left(\left(B4 - \text{``}B3\text{''}\right)\right)/\text{Float}\left(\left(\text{``}B4\text{''} + \text{``}B3\text{''}\right)\right) \tag{4.6}$$

Equation (4.5) was developed from a similar study conducted in Central Kalimantan, Indonesia, where NDVI is used as an indicator of vegetation health and vigour (Sugiarto & June, 2008) using the following equation:

$$\text{fAPAR} = 1.25 \times \text{NDVI} - 0.025 \tag{4.7}$$

The estimation of fAPAR using the NDVI-derived equation is critical, which is a remote sensing-based approach and is dependent on the availability of satellite data. We chose this equation because it can be used to indicate solar radiation energy that is used in the photosynthetic process, since fAPAR is part of PAR that is absorbed by plant canopy (Sugiarto & June, 2008).

4.5 Results and Discussion

4.5.1 Forest Density Mapping

In the present study, the vegetation density of the forest in Sungai Teliang in UMFR is characterized by various vegetation classes of forest, river and sedimentation percentage by area. The presence of sedimentation was only found in finer and moderate scales. In general, all scales classified water and non-water features were based on the SVM method.

The recorded forest density at the finer scales for forest, river and sedimentation was 97.9%, 1.30% and 0.70%, respectively. At moderate scale, forest, river and sedimentation was 52.8%, 46.6% and 0.70%, respectively. For coarse level at

Table 4.2 Forest density percentage by area for Sg. Teliang, Ulu Muda Forest Reserve

Method	Scale, level	Class	Forest density (%)
1.	50, finer	Forest	97.9
		River	1.30
		Sedimentation	0.70
2.	200, moderate	Forest	52.8
		River	46.6
		Sedimentation	0.70
3.	350, coarse	Forest	21.4
		River	78.6
		Sedimentation	No available

350 scale, forest density for forest and river was 21.4% and 78.6%, while the non-vegetation feature was also identified (Table 4.2).

In this study, the density mapping produced two vegetation classes and a land feature that consists of water and soil characteristics known as sedimentation. During the dry season, the water level in the river is reduced and, hence, exposed the ground river elements. Forest and river recorded good clusters at the finer scale. Meanwhile, forest and river in moderate scale showed clustering balance in discriminating between vegetation and non-vegetated features.

4.5.2 Forest Density Distribution Based on Area

The density distribution throughout the study recorded a range between less than 1 and 8.5 ha in all scales. At the finer scale, more attributes appeared for less than 1–2.5 ha. At moderate scale, more attributes were recorded from 6.0 to 7.5 ha. However, less polygons were recorded for 7.0–8.5 ha (Fig. 4.7). The results also showed that finer scale detected most of the pixel's spectral features identified within similar forest types in the UMFR.

4.5.3 Forest Density Distribution More than 1 ha

The density distribution was compared to low-scale attributes in areas of larger than 1 ha area with moderate and coarse scale attributes. The result showed that most of the areas were covered with more than 1 ha attributes (Fig. 4.8). The application of finer scale permits more attributes to be developed, resulting in more forest, land and water features being captured. Meanwhile, coarse scale captured less attributes within a larger area, which were mostly attributed to forest class.

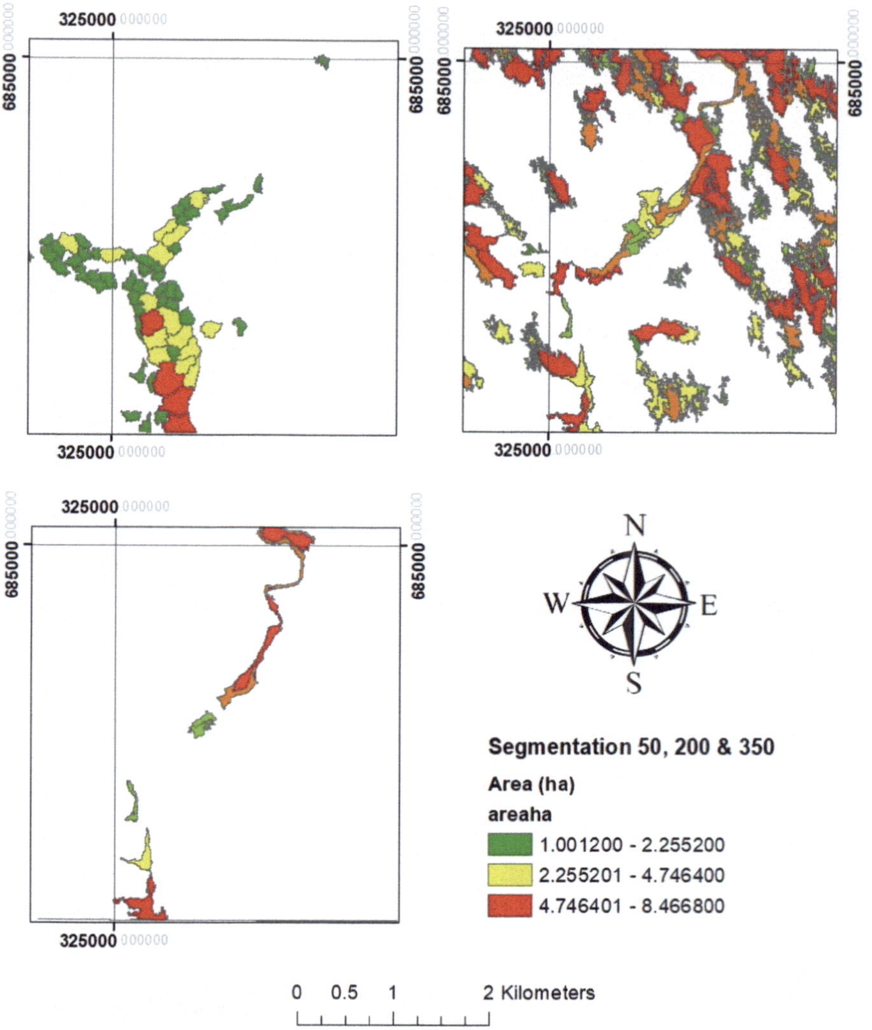

Fig. 4.7 Segmentation for the three methods

4.5.4 Accuracy Assessment

Matrices developed from the accuracy assessment procedures were tabulated as shown below. Because there were more attributes in finer scale, more polygons were selected for accuracy assessment analysis for forest area. As the scale number increases, fewer samples can be collected as accuracy samples (Tables 4.3, 4.4, and 4.5).

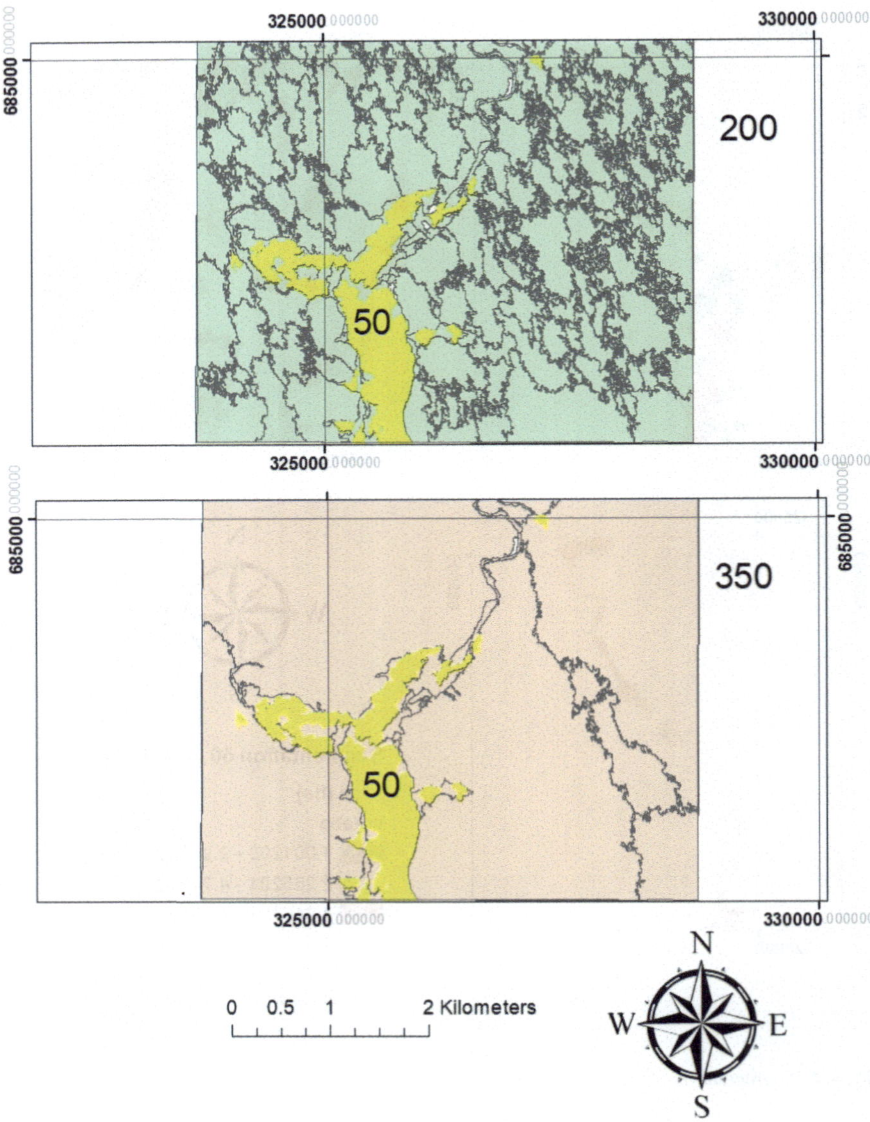

Fig. 4.8 Segmentation attributes for more than 1 ha area

4.5.5 Producer's and User's Accuracy

Based on accuracy assessment, the coarse scale showed a low producer's accuracy, which is 50%, owing to some of the forest being classified as a river. This is because the reflectance of the forest area is closer to sedimentation or elements in the river. However, because the user understood that the attributes should be classified as

Table 4.3 Error matrix for 50 scale

Classified class	Forest	River	Sedimentation	Total (user)
Forest	368	2	6	376
River	0	3	2	5
Sedimentation	0	2	0	2
Unknown	0	0	0	0
Total (producers)	368	7	8	383

Table 4.4 Error matrix for 200 scale

Classified class	Forest1	Sedimentation	River	Total (user)
Forest1	18	0	1	19
Sedimentation	0	0	0	0
River	0	5	9	14
Unknown	0	0	0	0
Total (producers)	18	5	10	33

Table 4.5 Error matrix for 350 scale

Classified class	Forest	River	Sedimentation	Total (user)
Forest	1	0	0	1
River	1	1	1	3
Sedimentation	0	0	0	0
Unknown	0	0	0	0
Total (producers)	2	1	1	4

forest, the user's accuracy was recorded as 100%. Non-forest features were not identified by this scale due to its roughness. The overall kappa for the coarse scale map was 1.0, while the opposite overall kappa was very low with a value of 0.27.

River attributes developed from moderate scale tend to confuse the user, and, in this study, user's accuracy was 94.74%, which is high, while the producer's accuracy was 100%. Producer has better attributes sampling for classifying river than the user, where some of the attributes from the forest were classified as river. However, the misclassification accuracy was higher due to the large number of attributed segments from this scale. Although river sampling collection was more difficult than coarse scale, the classification had a better producer's accuracy of 90% and lower user's accuracy of 64.29%. Kappa for forest was 0.88, while river was 0.49. This makes the overall kappa of 0.67 (Table 4.6).

Finer scale classification produced more attributes that makes it meaningful for sampling collection for the accuracy assessment. The map clearly showed that assigning river and sedimentation sampling is more confusing than in moderate scale, due to the too many attributions in both land covers. Forest provides harder assignments for user in sampling collection than for computer or producer to conduct classification. Therefore, the recorded producer's accuracy was 100%, whereas user's accuracy was 97.87%. Overall kappa for the scale was 0.46, while river's

Table 4.6 Accuracy assessment for final land use[a]

Method	Scale	Class	Producer's accuracy	User's accuracy	Overall accuracy (each scale)	Overall kappa	Overall kappa
1	50, finer	Forest	100.00	97.87	96.87	0.46	0.44
		River	42.86	60.00		0.59	
		Sedimentation	0.00	0.00		– 0.02	
2	200, moderate	Forest	100.00	94.74	81.82	0.88	0.67
		River	90.00	64.29		0.49	
		Sedimentation	0.00	0.00		0.00	
3	350, coarse	Forest	50.00	100.00	50.00	1.00	0.27
		River	100.00	33.33		0.11	
		Sedimentation	0.00	0.00		0.00	

[a] Note: Shape, compactness = 0.5, 0.5

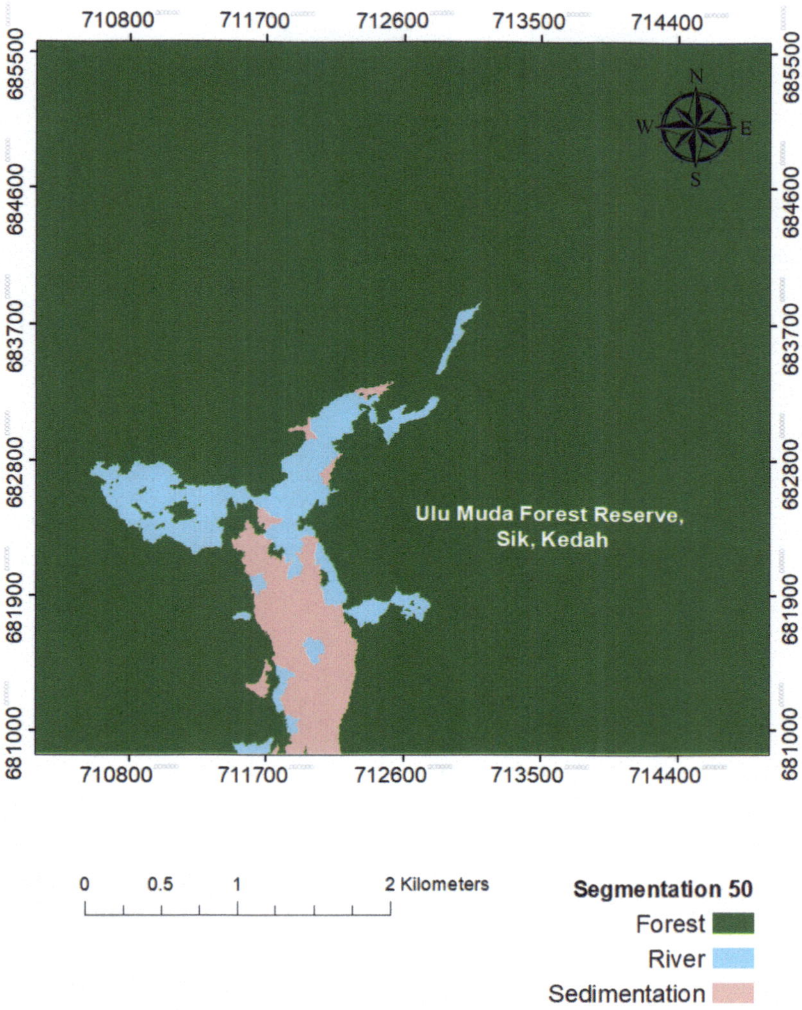

Fig. 4.9 Final land use map for segmentation—50 scale

overall kappa was higher than forest's 0.59. Moderate scale map showed the best kappa with 0.67 value.

During the dry season, sedimentation occurs as a result of the drying of water in the river. Sedimentation appears in two of the earlier scales but disappears in coarse scale in land use map. The study found that sedimentation class was not counted in the accuracy assessment due to sample inadequacy as determined by the computer algorithm. The final land use map is shown in Figs. 4.9, 4.10 and 4.11.

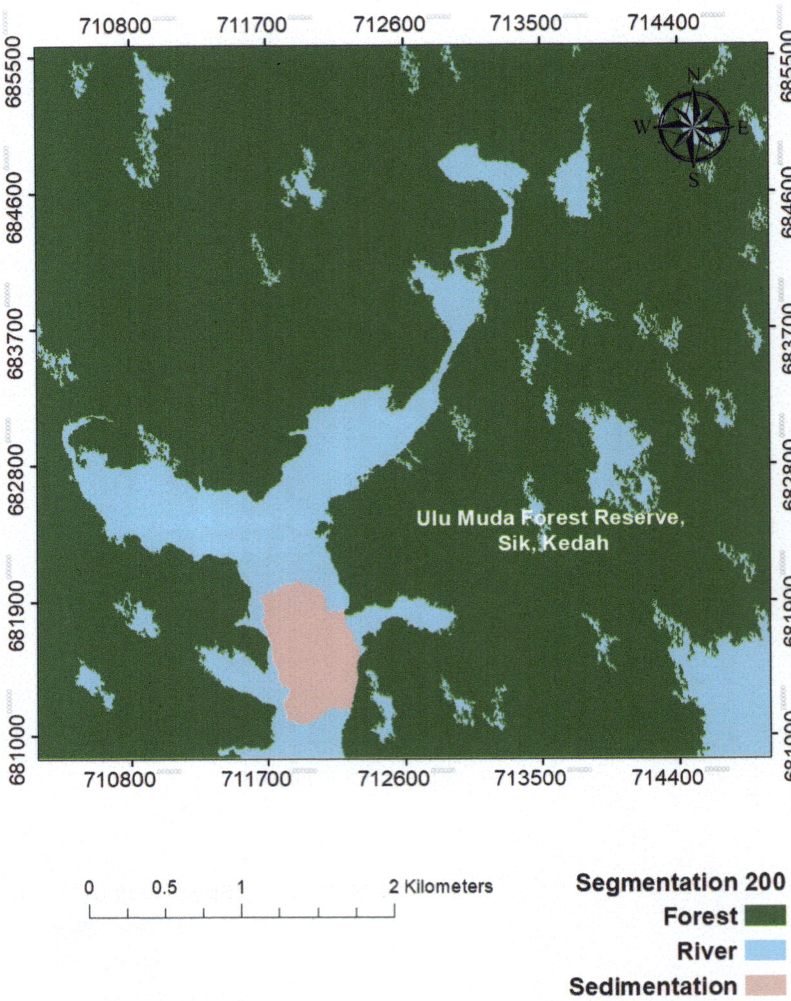

Fig. 4.10 Final land use map for segmentation—200 scale

4.5.6 Biophysical and Vegetation Indices

Understanding the statistics of NDVI is essential for studying vegetation. Here, the study derived NDVI for all the segmentation methods (Fig. 4.12). Generally, the reported NDVI values in other studies vary with different types of land cover types. For example, a study by Gebremichael and Barros (2006) used NDVI maximum and minimum values to derive fractional vegetation cover (*fv*).

In this study, the NDVI mean value was derived for all the segmentation methods. Mean NDVI was 0.709, which is the highest for all the methods, whereas the mean for finer and moderate scales was 0.696 and 0.698, respectively. This finding showed

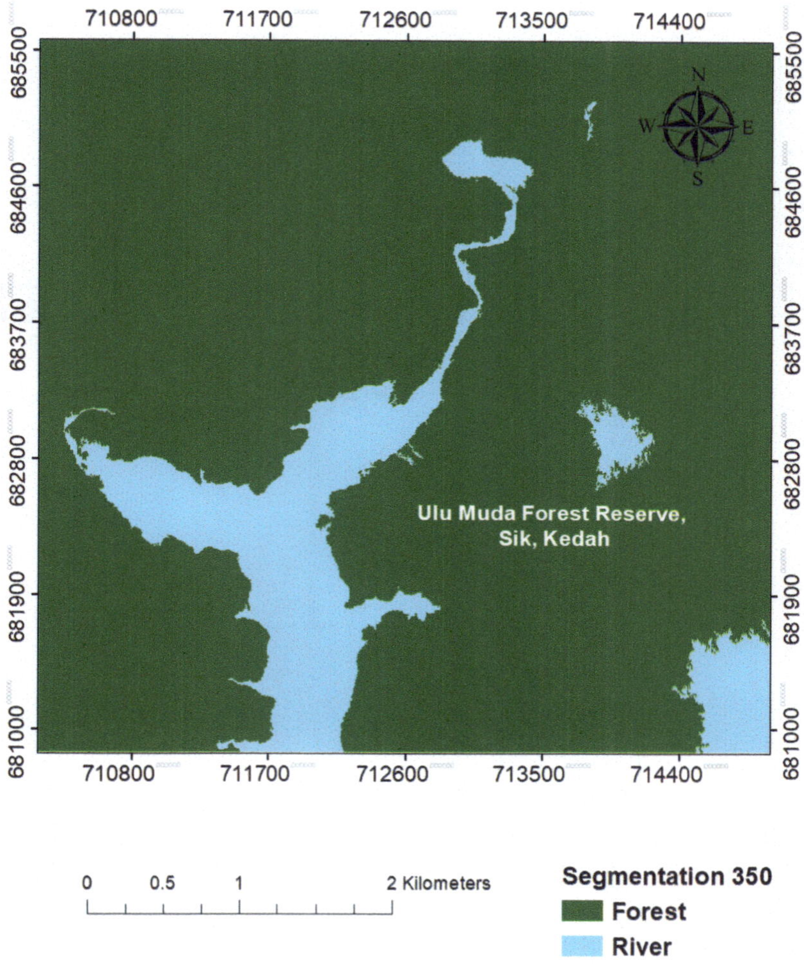

Fig. 4.11 Final land use map for—350 scale

that lower NDVI was caused by the higher spectral reflectance in river in the presence of a large number of forest attributes in the coarse method. This finding can also be associated with higher NDVI as a result of the high biomass area since NDVI is an indicator of biomass level, as demonstrated in the mangrove area in Maubesi Nature Reserve in Indonesia (Pujiono et al., 2013). NDVI is best known as a useful method for distinguishing between vegetation species and assessing the health of vegetation (Luus & Kelly, 2008). In addition, recent study supported that utilization of NDVI as vegetation index (VI) is proven as a good indicator in assessing phenology which is at the meantime a good input for NPP estimation (Maleki et al., 2020).

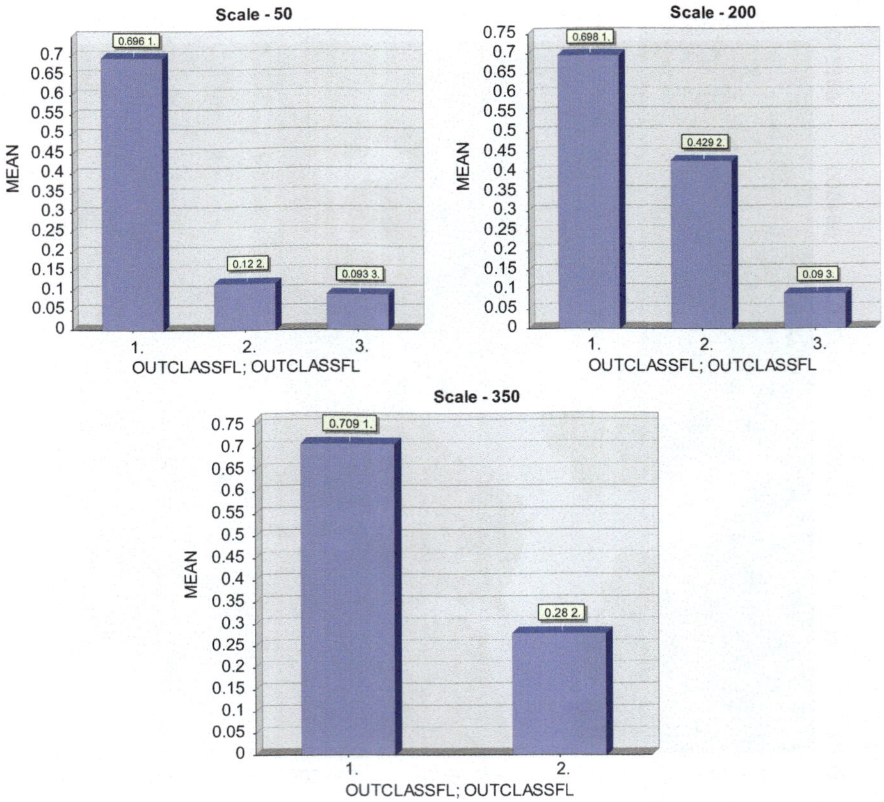

Fig. 4.12 Final land use map for (**a**) 50 (upper-left), (**b**) 200 (upper-right) and (**c**) 350 (lower). (Note: 1, 2 and 3 shown in the graph are forest, river and sedimentation class, respectively)

4.5.7 NPP

The NPP for the study area ranged from 6.7 g C m^{-2} $month^{-1}$ to a maximum value of 300.04 g C m^{-2} $month^{-1}$ (Fig. 4.13). In general, these results show that the UMFR has high forest biomass and that it is an intact forest that has been logged with very good forest management and logging practices.

This is because logging operations have been known to highly influence the environment of tropical forests (Sadeghi et al., 2017). Since 1901, the forest management at UMFR adhered to the Sustainable Forest Management (SFM) on logging operations (MTC, 2020). Through the practice of Selective Management System (SMS) in the area, it has evolved to optimize an economic cut to ensure the sustainability of the forests and the lowest cost for forest development. Therefore, there will be sufficient rehabilitation activities to cover the forest ground and protect it from the effects of forest operations. Another study reported that the NPP in Pasoh

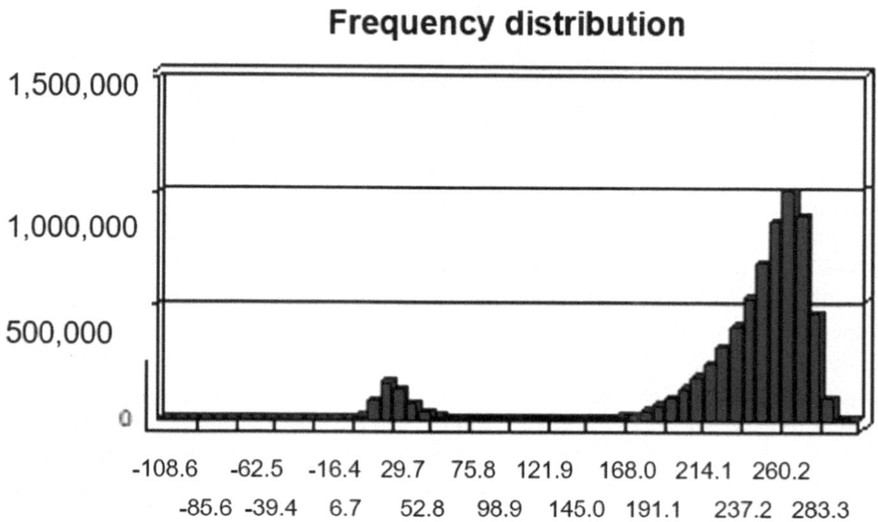

Fig. 4.13 Minimum and maximum value of NPP for the study

Forest Reserve ranges between 202.2 and 133.4 g C m^{-2} $month^{-1}$ (Razali et al., 2015), which corresponds to the same forest features as the current study.

In particular for the case of tropical forest, interannual variation of NPP in the region of tropical rainforest forest of Queensland (Shiba & Apan, 2011) seems to be influenced by rainfall amount, maximum temperature, solar radiation and potential evaporation factors. Hence, the understanding of NPP may enhanced forest management practices for formulation of fertilization regime during dry and wet season. NPP can be indicator to forest fertility because there will be another environmental factor interacting with these factors in the processes of driving NPP in the region. This is because more resources, namely, researchers, tools and specific instruments, are required to assess directly the influence of nutrients, soil pH, cloud cover and soil type (Shiba & Apan, 2011). By deriving NPP values from remote sensing, it showed relation between NPP and climatic condition (Indiarto & Sulistyawati, 2013).

It is important to assess NPP for tropical forest because this forest plays a huge role in slowing down the rate of global warming by about 15% (Malhi, 2010). Early study showed that areas in Malay Peninsula, Sumatra, Borneo, Celebes and north-western of New Guinea showed NPP in declining rate from 2010 to 2013 due to deforestation (Shao & Zeng, 2017); in fact the study also suggests it also could be due to forest degradation. NPP decreases dramatically after natural forest is logged for forest plantation due to the removal of forest cover on the ground. However, as plantations began to cover the land, the respiratory rate in this warm climate induces a low NPP/GPP ratio rate in the tropical region. Simultaneously, fertilizer application during the rainy season caused more nutrient deficiency, which may result in a low NPP value. Therefore, NPP is a good indicator of good and bad forest plantation management practices in a specific area.

4.5.8 NPP and Its Impact on the Ulu Muda FR Community

Full-time fishermen, subsidy recipient and non-subsidy recipient, part-time or seasonal basis, were identified among the respondents interviewed. According to the study, the majority of respondents were between the ages of 41 and 60, with the remainder falling between the ages of 61 and 80. Meanwhile, respondents had varying years of experience as fisherman, which included only about 9 years (17.4%), 11–20 years of experience (23.9%), 21–30 years (26.1%), 31–40 years (23.9%) and more than 40 years of experience (8.7%).

Although the extraction process of fishing uses biomass, which is powered by solar energy, the amount is not calculated in this study NPP analysis. However, the study assumed that the process is part of their socio-economic process, which has an impact on the forest. The NPP value for the study was higher than in other areas with similar forest types. First, the study assumed that their activities were not harmful to the forest and that no major NPP were extracted from the forest in order to carry out their activities. Meanwhile, the following people received subsidies: Kuala Muda (33) recipient, Sik (12), Kulim (9) recipient and Baling (4) recipient. The majority of those polled considered themselves to be subsidy recipients. Subsidies can lower the cost of fishing operations while increasing revenue, making fishing businesses more profitable (Ali et al., 2017).

4.6 Conclusion

Forest density and NPP mapping based on remote sensing satellite image of GeoEye-1 are some of the approaches that can be used to estimate the NDVI as biophysical indices in estimating NPP at UMFR, particularly for Sungai Teliang. Remote sensing based on high-resolution images plays a huge role as potential data that should be used as a reference to other NPP studies in forest evergreen areas, particularly in Malaysian Peninsular. In this study, mapping forest density using OBIA has limited potential, while other techniques should be explored. Nevertheless, OBIA can be used with high-resolution data to produce a more meaningful and accurate assessment. The NPP results from this study support earlier findings that characterize the forest area as having high biomass. Moreover, the study showed NPP value is higher than other area with similar forest types and no major NPP extracted by the Ulu Muda FR fisherman community assessed from the study. The study also showed that forest density can be mapped using remote sensing technology, whereby final land use map is also created from that analysis.

Acknowledgements The authors would like to thank WWF-Malaysia for funding this project, even though this paper reports only a part of the project implementation. Gratitude is also extended to the Institute of Tropical Forestry and Forest Products (INTROP), Universiti Putra Malaysia, for funding for software and data acquisition for this study.

References

Abdullah, W. S. W., Osman, M., Kadir, M. Z. A. A., & Verayiah, R. (2019). The potential and status of renewable energy development in Malaysia. *Energies, 12*(12), 2437. https://doi.org/10.3390/en12122437

Aik, Y. C., Lim, K. C., & Hymeir, K. (2017). *Surveys and monitoring of the plain-pouched Hornbills in The Greater Ulu Muda Forest Complex, Kedah Darul Aman, Peninsular Malaysia (2012–2013)*. https://doi.org/10.13140/RG.2.2.13492.86409

Ali, J., Abdullah, H., Saifoul, M., Noor, Z., Kuperan Viswanathan, K., & Islam, G. N. (2017). The contribution of subsidies on the welfare of fishing communities in Malaysia. *International Journal of Economics and Financial, 7*(2), 641–648. http://www.econjournals.com

Band, L. E., Csillag, F., Ferera, A. H., & Baker, J. A. (1999). *Deriving an eco-regional framework for Ontario through large-scale estimates of net primary productivity*. Ontario Forest Research Institute.

Chew, M. Y., Hymeir, K., Nosrat, R., & Shahfiz, M. A. (2014). Relation between grasses and large herbivores at the Ulu Muda Salt Licks, Peninsular Malaysia. *Journal of Tropical Forest Science, 26*(4), 554–559.

Cohen, J. (1960). A coefficient of agreement for nominal scales. *Educational and Psychological Measurement, 20*(1), 37–46.

Coops, N. C., Hilker, T., Hall, F. G., Nichol, C. J., & Drolet, G. G. (2010). Estimation of light-use efficiency of terrestrial ecosystems from space: A status report. *BioScience, 60*(10), 788–797. https://doi.org/10.1525/bio.2010.60.10.5

Faidi, M. A., Ibrahim, A. L., Wahid, A. R., & Huey, T. (2010). The capability of eco-physiological approach to determine net primary productivity (NPP) of tropical rainforest using remote sensing data. In *31st Asian Conference on Remote Sensing 2010, ACRS 2010*, 727–733.

Gebremichael, M., & Barros, A. (2006). Evaluation of MODIS Gross Primary Productivity (GPP) in tropical monsoon regions. *Remote Sensing of Environment, 100*(2), 150–166. https://doi.org/10.1016/j.rse.2005.10.009

Girardin, C. A. J., Malhi, Y., Aragão, L. E. O. C., Mamani, M., Huaraca Huasco, W., Durand, L., Feeley, K. J., Rapp, J., Silva-Espejo, J. E., Silman, M., Salinas, N., & Whittaker, R. J. (2010). Net primary productivity allocation and cycling of carbon along a tropical forest elevational transect in the Peruvian Andes. *Global Change Biology, 16*(12), 3176–3192. https://doi.org/10.1111/j.1365-2486.2010.02235.x

Goward, S. N., Waring, R. H., Dye, D. G., & Yang, J. (1994). Ecological remote sensing at OTTER: Satellite macroscale observations. *Ecological Applications, 4*, 322–343.

Handcock, R. N., & Csillag, F. (2004). Spatio-temporal analysis using a multiscale hierarchical ecoregionalization. *Photogrammetric Engineering & Remote Sensing, 70*(1), 101–110.

Hashimoto, H., Wang, W., Milesi, C., White, M. A., Ganguly, S., Gamo, M., Hirata, R., Myneni, R. B., & Nemani, R. R. (2012). Exploring simple algorithms for estimating gross primary production in forested areas from satellite data. *Remote Sensing, 4*(12), 303–326. https://doi.org/10.3390/rs4010303

Indiarto, D., & Sulistyawati, E. (2013). Monitoring net primary productivity dynamics in Java Island using MODIS satellite imagery. *34th Asian Conference on Remote Sensing 2013, ACRS 2013, 2*, 1730–1737.

Kimball, J. S., Zhao, M., McDonald, K. C., & Running, S. W. (2006). Satellite remote sensing of terrestrial net primary production for the Pan-Arctic Basin and Alaska. *Mitigation and Adaptation Strategies for Global Change, 11*(4), 783–804. https://doi.org/10.1007/s11027-005-9014-5

Luus, K. A., & Kelly, R. E. J. (2008). Assessing productivity of vegetation in the Amazon using remote sensing and modelling. *Progress in Physical Geography, 32*(4), 363–377. https://doi.org/10.1177/0309133308097029

Malaysian Meteorological Department. (2018). *Precipitation data for Ulu Muda Dam, Sik, Kedah (1989–2018)*. Malaysian Meteorological Department.

Maleki, M., Arriga, N., Barrios, M., Wieneke, S., Liu, Q., Peñuelas, J., Janssens, I. A., & Balzarolo, M. (2020). Estimation of gross primary productivity (GPP) phenology of a short-rotation plantation using remotely sensed indices derived from sentinel-2 images. *Remote Sensing, 12*(2104), 1–26.

Malhi, Y. (2010). The carbon balance of tropical forest regions, 1990–2005. *Current Opinion in Environmental Sustainability, 2*(4), 237–244. https://doi.org/10.1016/j.cosust.2010.08.002

Mei, S. L. (2020). *Socio-economic and livelihood assessment of Inland fishermen in Muda river basin* (pp. 586–594). https://doi.org/10.15405/epsbs.2020.10.02.53

Miniandi, N. D., Mohd Sofiyan, S., Razak, Z., Sheriza, M. R., & Nor Rohaizah, J. (2021). Morphometric analysis of the main river systems at Ulu Muda Forest Reserve, Kedah, Peninsular Malaysia. *The Malayan Nature Journal, 73*(3), 297–309.

MTC. (2020). *Resource forest and sustainability: Sustainable forestry in Malaysia*. Malaysian Timber Council. Retrieved from http://mtc.com.my/resources-SustainableForestryinMalaysia.php

Muhamad, L. (2010, August). *Estimating net primary productivity (NPP) on tropical rain forest using MODIS observation data*. https://doi.org/10.13140/RG.2.1.2954.2249

Muzathik, A. M., Mohd, W., Bin, N., Nik, W., & Terengganu, K. (2010). Reference solar radiation year and some climatology aspects of East Coast of West Malaysia. *American Journal of Engineering and Applied Sciences, 3*(2), 293–299.

Pachavo, G., & Murwira, A. (2014). Remote sensing net primary productivity (NPP) estimation with the aid of GIS modelled shortwave radiation (SWR) in a Southern African Savanna. *International Journal of Applied Earth Observation and Geoinformation, 30*, 217–226. https://doi.org/10.1016/j.jag.2014.02.007

Peng, C., & Apps, M. J. (1999). Modelling the response of net primary productivity (NPP) of boreal forest ecosystems to changes in climate and fire disturbance regimes. *Ecological Modelling, 122*(3), 175–193. https://doi.org/10.1016/S0304-3800(99)00137-4

Pujiono, E., Kwak, D. A., Lee, W. K., Sulistyanto, D., Kim, S. R., Lee, J. Y., Lee, S. H., Park, T., & Kim, M. I. (2013). RGB-NDVI color composites for monitoring the change in mangrove area at the Maubesi Nature Reserve, Indonesia. *Forest Science and Technology, 9*(4), 171–179. https://doi.org/10.1080/21580103.2013.842327

Rasib, A. W., Ibrahim, A. L., Cracknell, A. P., & Faidi, M. A. (2008). Local scale mapping of net primary production in tropical rain forest using MODIS satellite data. In *The International Archives of the Photogrammetry, Remote Sensing and Spatial Information Science, XXXVII (Part B7)* (pp. 1441–1446).

Razali, S. M., Atucha, A. A. M., Nuruddin, A. A., Shafri, H. Z. M., & Hamid, H. A. (2015). Mapping human impact on net primary productivity using MODIS data for better policy making. *Applied Spatial Analysis and Policy, 9*(3), 389–411. https://doi.org/10.1007/s12061-015-9156-0

Rouse, J. W., Hass, R. H., Schell, J. A., & Deering, D. W. (1973). Monitoring vegetation systems in the great plains with ERTS. *Third Earth Resources Technology Satellite (ERTS) Symposium, 1*, 309–317.

Sadeghi, S. M., Faridah-Hanum, I., Wan Razali, W. M., Abd Kudus, K., & Hakeem, K. R. (2017). Tree composition and diversity of a hill dipterocarp forest after logging. *Malayan Nature Journal, 66*(4), 1–15.

Satellite Imaging Corporation. (2017). *GeoEye-1 satellite sensor*. Retrieved from https://www.satimagingcorp.com/satellite-sensors/geoeye-1/

Senna, M. C. A. (2005). Fraction of photosynthetically active radiation absorbed by Amazon tropical forest: A comparison of field measurements, modeling, and remote sensing. *Journal of Geophysical Research, 110*(G1), G01008. https://doi.org/10.1029/2004JG000005

Shao, P., & Zeng, X. (2017). Comparative analysis of NPP changes in global tropical forests from 2001 to 2013. *IOP Conference Series: Earth and Environmental Science, 57*(1), 012009. https://doi.org/10.1088/1742-6596/755/1/011001

Sharma, D. S. K., Lee, B. M. S., & W., A. Z. A., & S., S. (2005). Rapid assessment of terrestrial vertebrates in Sg. Lasor, Ulu Muda Forest Reserve. In *Hutan Simpan Ulu Muda, Kedah: Pengurusan, Persekitaran Fizikal dan Biologi* (pp. 212–221). Jabatan Perhutanan Semenanjung Malaysia.

Shiba, S. S. T., & Apan, A. (2011). Analysing the effect of drought on net primary productivity of tropical rainforests in Queensland using MODIS satellite imagery. In *Proceedings of the Surveying & Spatial Sciences Biennial Conference 2011 21–25 November 2011, Wellington, New Zealand* (pp. 97–110).

Sugiarto, Y., & June, T. (2008). Estimation of net primary production (NPP) using remote sensing approach and plant physiological modeling. *Agromet, 22*(2), 183–199. https://doi.org/10.29244/j.agromet.22.2.183-199

Sunaryathy, P. I., Busu, I., & Rasib, A. W. (2012). Evaluation of net primary productivity of oil palm plantation in South Sulawesi Indonesia. In *The 33rd Asian Conference on Remote Sensing* (p. 5).

Turner, D., Ritts, W., & Cohen, W. (2006). Evaluation of MODIS NPP and GPP products across multiple biomes. *Remote Sensing of Environment, 102*, 282–292.

Vu, Q. M., Le, Q. B., & Vlek, P. L. G. (2014). Hotspots of human-induced biomass productivity decline and their social–ecological types toward supporting national policy and local studies on combating land degradation. *Global and Planetary Change, 121*, 64–77. https://doi.org/10.1016/j.gloplacha.2014.07.007

Xiao, X., Hollinger, D., Aber, J., Goltz, M., Davidson, E. A., Zhang, Q., & Moore, B. (2004). Satellite-based modeling of gross primary production in an evergreen needleleaf forest. *Remote Sensing of Environment, 89*, 519–534. https://doi.org/10.1016/j.rse.2003.11.008

Zhang, R., Zhou, Y., Luo, H., Wang, F., & Wang, S. (2017). Estimation and analysis of spatiotemporal dynamics of the net primary productivity integrating efficiency model with process model in karst area. *Remote Sensing, 9*(5), 477. https://doi.org/10.3390/rs9050477

Zhang, K., Liu, N., Chen, Y., & Gao, S. (2019). Comparison of different machine learning method for GPP estimation using remote sensing data. *IOP Conference Series Materials Science and Engineering, 490*(6), 062010. https://doi.org/10.1088/1757-899X/490/6/062010

Temporal Reduction of Forest Ecosystem Services and Drivers of Deforestation

5

Mahirah Kamaludin and Moe Shwe Sin

Abstract

The tangible and intangible benefits of the forest ecosystem need to be preserved, assessed and seriously considered by policymakers especially under the present threat, which is the extinction of forest ecosystem services in Malaysia. This happened due to uncontrolled deforestation and to fulfil the increasing demand by population. The direct and indirect determinants of deforestation are important in determining the long-run impacts of deforestation in Malaysia, especially in the urban area. In this case, the empirical study was carried out using the autoregressive distributed lag (ARDL) approach in a study site in Peninsular Malaysia. It was found that the total population, urban population, agricultural land and gross domestic product (GDP) are the direct and indirect determinants that have a significant, long-run association between total forest area and deforestation. The temporal reduction of total forest area leads to the failure of ecosystem services and may affect the country's sustainability and social living.

Keywords

Deforestation · Forest ecosystem · ARDL model

M. Kamaludin · M. S. Sin (✉)
Faculty of Business, Economics and Social Development, Universiti Malaysia Terengganu, Kuala Nerus, Terengganu, Malaysia
e-mail: moe.sin@umt.edu.my

Z. Samdin et al. (eds.), *Tropical Forest Ecosystem Services in Improving Livelihoods For Local Communities*, https://doi.org/10.1007/978-981-19-3342-4_5

5.1 Introduction

Forest ecosystems around the world deliver various services, and they are vital for the survival of mankind. A forest stipulates biodiversity maintenance and conservation by serving as the principal habitat for a diverse range of flora and fauna. Forest ecosystems offer services on a large scale and in such that the technology cannot effectually replace them. There are four categories of ecosystem services, i.e. (1) provisioning services or the provision of food or habitat; (2) regulating services, e.g. the regulation of erosion or climate; (3) supporting services, e.g. primary production or nutrient cycling; and (4) cultural services, e.g. aesthetic enjoyment or recreation (Millennium Ecosystem Assessment Panel, 2005). Figure 5.1 demonstrates various combinations of ecosystem services offered by a healthy forest. The categorization is important to better understand the potential applications of ecosystem services and postulate a framework to analyse both active and passive influences that ecosystem services have on human well-being.

Forests give benefits to human well-being in various ways aside from basic survival. Generally, a healthy forest stimulates healthy environment and surroundings. For instance, the trees lower the stormwater runoff, increase the soil's ability to absorb and store water as well as avoid soil erosion, all of which contribute to healthier and cleaner waterways and lesser flood. However, the recent decades have shown a dramatic increase in the human population whose non-traditional agricultural and cultivation activities have expanded to exploiting unsuitable lands (Ishizuka et al., 2000; Ying et al., 2018; Hattori et al., 2019). These activities are indeed unsustainable because farmers burn the forests to have a short-period cultivation process (Tanaka et al., 2005; Ying et al., 2018) and can cause severe forest degradation characterized by soil erosion, loss of soil fertility and decreasing biomass and species richness (Kleinman, 1995; Lamb, 2010; Hepp et al., 2018; Ying et al., 2018).

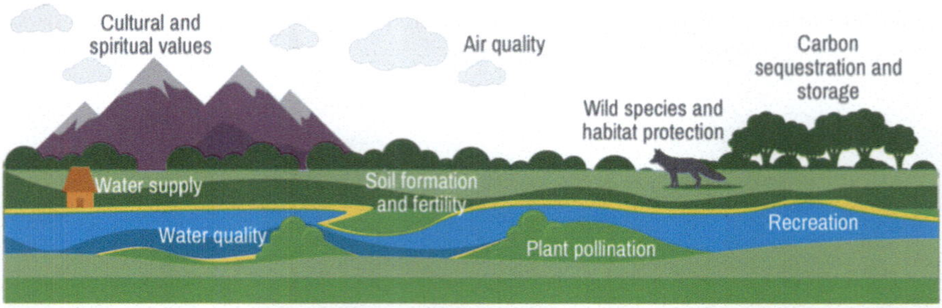

Fig. 5.1 Ecosystem services provided by healthy forest landscapes. (Source: Forest Trends' Ecosystem Marketplace (2016). *An Atlas of Ecosystem Markets in the United States.* Washington DC: Forest Trends)

5.1.1 Deforestation

Hancock (2021) calls for deforestation and forest degradation as urgent matters on a global scale. About 47% of the world's forests are at high risk for deforestation or degradation by 2030 (United Nations, 2021). While both are damaging to forest health, there is a difference between deforestation and forest degradation. Deforestation refers to the clearing of a forest, entirely removing it to put up something else in its place. The main driver of deforestation is unsustainable agriculture such as palm oil and rubber. When a forest is degraded, it still exists, but it can no longer function well. Forest health declines until it can no longer support people, wildlife and surroundings. Forest degradation is an even bigger problem than deforestation. About 6.5 million square miles of forest are at high risk of degradation in the next 10 years (FAO, 2016). In Malaysia from 2001 to 2019, about 95% of tree cover loss occurred in areas where the dominant drivers of loss resulted in deforestation as shown in Fig. 5.2 (GFW, Global Forest Watch, 2021).

The total deforested area was found at 200,225 ha between 2010 and 2015, accounting for deforestation at the rate of 40,045 ha per year (FAO, 2016) (Table 5.1). A large piece of forest areas in Peninsular Malaysia have been converted to oil palm and rubber at commercial scales since the 1980s, occupying around 61% and 14%, respectively (Omar & Misman, 2018), and in 2000 and 2012, the expansion of oil palm plantation occurred in Malaysia, and it threatened the biodiversity conservation and other negative environmental impacts (Shevade & Loboda, 2019).

It is important to note that degradation will prevent subsequent forest recovery and biomass accumulation (Chazdon, 2003; Villa et al., 2018). For instance, in tropical rainforests, the soil nutrient cycle changes drastically once the forest is

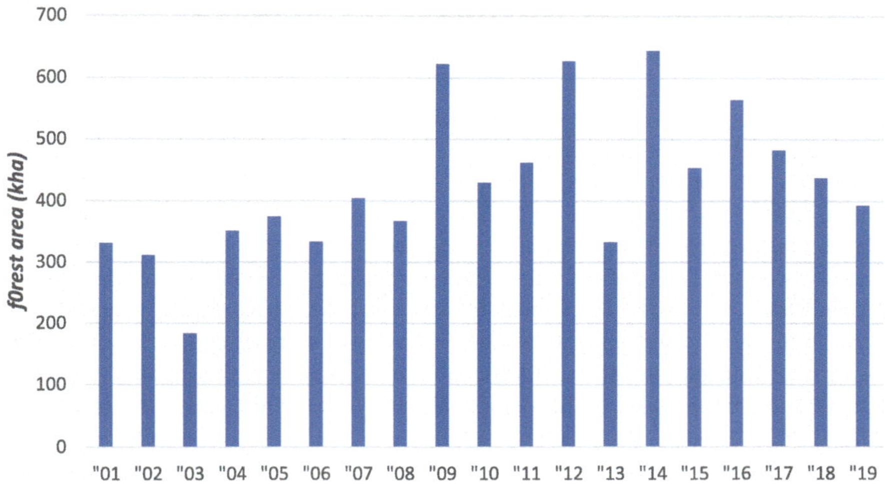

Fig. 5.2 Annual tree cover loss in Malaysia from 2001 until 2019. (Source: GFW, Global Forest Watch (2021))

Table 5.1 Deforestation in Peninsular Malaysia between 2010 and 2015

Forest type	Forest cover 2010 (ha)	Forest cover 2015 (ha)	Deforestation (ha)	Rate of deforestation (ha per year)	Rate of deforestation (% per year)
Inland	5,690,816	5,525,034	165,782	33,156	0.58
Peat swamp	290,038	264,578	25,460	5092	1.76
Mangrove	115,181	106,198	8982	1796	1.56
Total	6,096,035	5,895,810	200,225	40,045	0.66

burned or removed (Malmer et al., 1998; Giardina et al., 2000). Hence, the degradation of forest ecosystem services has led to many challenges and difficulties particularly because of anthropogenic activities. Commercial and artisanal logging, large-scale land conversion, fuelwood and charcoal production, slash and burn agriculture, non-timber forest product harvesting, hunting and mining all have impacts on forest biodiversity.

Malaysia's rate of deforestation is accelerating faster than any other tropical country in the world. Data from the United Nations, specifically the figure analysis from Food and Agriculture Organization (FAO), shows that Malaysia's annual rate of deforestation jumped almost 86% in the 2000–2005 period compared to the 1990–2000 period. The major cause of deforestation in Peninsular is highlighted as the expansion of agricultural plantations and extension of permanently cultivated land (Shevade & Loboda, 2019). In total, Malaysia has suffered a loss of an average of 140,200 ha per year since 2000. For comparison, the Southeast Asian countries lost an average of 78,500 ha annually during the 1990s.

5.1.2 Determinants of Deforestation and the Impact of Deforestation in Peninsular Malaysia

The rate of deforestation in the state of Selangor is currently the highest in Peninsular Malaysia and has rapidly grown since the past 20 years (Aisyah et al., 2015). It has lost 24% of its forest reserves, particularly in Semenyih and Hulu Langat areas as shown in Fig. 5.3, accounting for 57% of the total loss. Thirty years ago, in 1990, Semenyih, Selangor State, was an area occupied with rubber estates and has since transformed to agricultural activities when the demand for rubber decreased and the demand for crops increased (Global Forest Watch Malaysia, 2021). Floods often occur in this area, causing great damages to business premises and the communities losing their valuable properties. Additionally, in the history of this country, a total of RM200 million loss is one of the major consequences of deforestation (Hansen et al., 2015).

Understanding the determinants of deforestation and the impacts they have on deforestation, socio-economic development and environmental degradation is important. Deforestation activities in Malaysia usually involve the conversion of

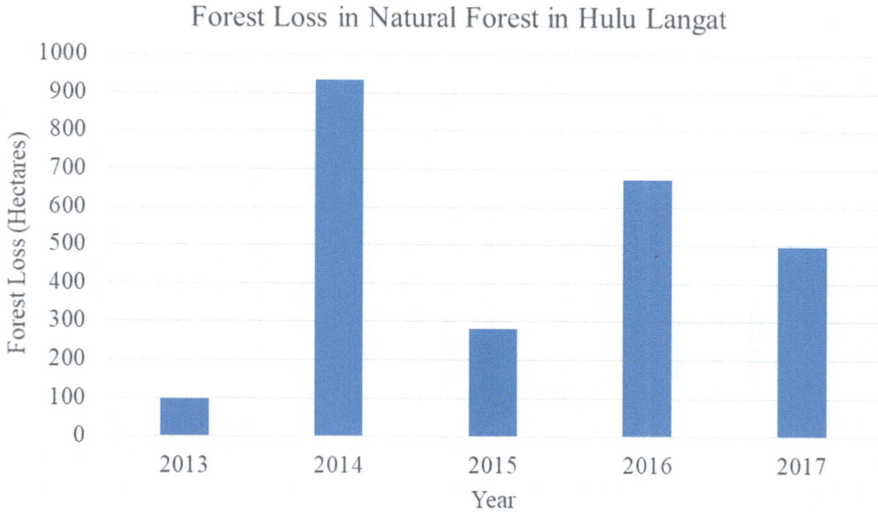

Fig. 5.3 Natural forest loss in Hulu Langat. (Source: Global Forest Watch)

forests to the plantation and urban areas (suggested to put citation for this statement). Direct determinants of deforestation are the first-level causes of forest loss which can easily be identified (Young et al., 2015; Austin et al., 2019) including clearing forests for agricultural and farming purposes. Ferronato and Torretta (2019) argued there is a positive relationship between the urbanization process and deforestation, in which forest clearing in city areas is an unavoidable process to solve the problem of increased population. Other indirect determinants such as low agricultural input price for the expansion of agricultural farms (Jellason et al., 2021), increasing credit availability and technological advancement also encourage deforestation (Assunção et al., 2019; Bologna & Aquino, 2020). Population density is another indirect determinant that increases the demand for residential land and buildings. In the present study, the magnitude of deforestation, which is driven by direct and indirect determinants, is analysed to call attention to the resilience and maintenance of forests in Peninsular Malaysia. The short- and long-run effects are analysed using econometric modelling on time series secondary data retrieved from published secondary data sources from World Bank and United Nations. The awareness of the importance of determinants of deforestation and the impacts of reducing forest areas is highlighted along with the implementation of policies for the country's economic growth.

5.1.3 Greenhouse Gas Emissions

Forests are more than just a collection of flora and fauna. They are integrated ecosystems, home to some of the most diverse lives on earth and the major players

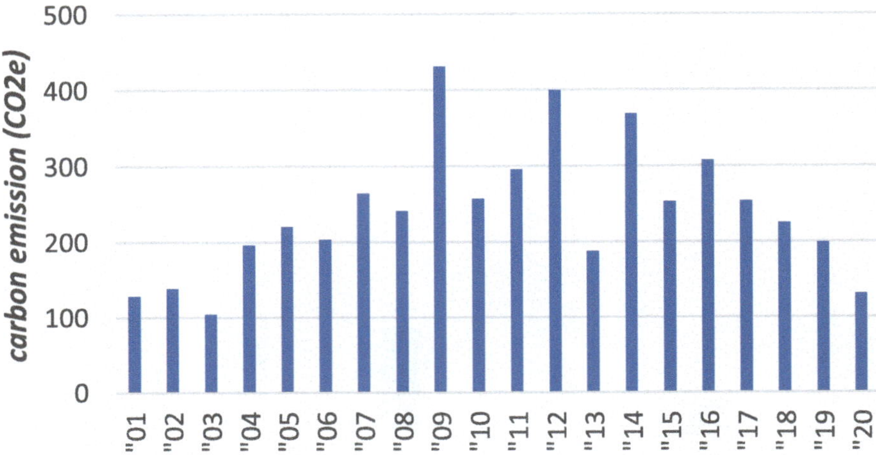

Fig. 5.4 Forest-related greenhouse gas emissions in Malaysia from 2001 until 2020. (Source: GFW, Global Forest Watch (2021))

in the carbon and water cycles. Forest degradation leads to a series of destructions and unfortunate changes that affect many lives. Forests function as carbon sinks which helps to mitigate the carbon dioxide (CO_2) emissions and other greenhouse gases, for instance, tropical forests alone hold more than 228–247 gigatons of carbon which is more than seven times the amount emitted each year by human activities (Pugh et al., 2019). Nevertheless, the loss of forests emits CO_2 instead of absorbing it. Thus far, deforestation and forest degradation are responsible for around 15% of all greenhouse gas emissions, contributing to rising temperature, changes in the patterns of weather and water supply and increased frequency of extreme weather events (WWF, 2021). Between 2001 and 2020, an average of 241 Mt per year was released into the atmosphere because of tree cover loss in Malaysia (GFW, Global Forest Watch, 2021). A total of 4.82 Gt of CO_2e was emitted during this period (Fig. 5.4).

5.2 Deforestation and Impact of Drivers in Selangor, Peninsular Malaysia

The association and effect of direct and indirect drivers on deforestation in Selangor State, Peninsular Malaysia, were analysed by using time series secondary data extracted from World Bank data source and United Nations data. The association of direct drivers such as total agricultural land in Selangor State and indirect drivers such as urban population, total population and GDP per capita on total forest area in Selangor State was analysed with econometric modelling using EView software. The specification of the model was discussed in session 5.2.2.

5.2.1 Study Area and Conceptual Framework of Deforestation

Semenyih, a town and district located in the district of Hulu Langat in southeastern Selangor, was selected as the study area. Selangor, a state with the highest gross domestic product (GDP) (DOSM, 2020) situated on the west coast of Peninsular Malaysia, occupies 8104 km^2 of land. It is the most populous state in the country (20.1%), followed by Sabah (11.7%) and Johor (11.6%) out of the total population of 32.7 million in Malaysia (Department of Statistics Malaysia, 2021). Semenyih is famous with Tekala Forest Reserve which is located in Sungai Lalang Forest Reserve, a lowland dipterocarp forest. Figure 5.5 shows the map of state of Selangor, Malaysia.

The major environmental issue in Selangor is the loss of total forest areas year by year because of the expansion of agricultural land for oil palm plantation, causing the flora and fauna to suffer from losing their habitats (Shevade et al., 2017). Forest reserves in Semenyih have transformed into high-end residential housing areas. However, Bhatia and Cumming (2020) argued that as the society becomes wealthier, the destruction of forests will cause the society to become more aware of environmental issues and will be more willing to invest in environmental protection. However, there might be limited forest transition theory in the sense of treatment of forest dynamics and transition (Christian, 2017). The dynamic transition of forest in terms of total forest area is mainly concerned with economic growth factors in the short and long term. The conceptual framework of deforestation in terms of total forest area with the impact of its determinants is shown in Fig. 5.6.

5.2.2 Hypothesis of the Study and Specification of the Model

The relationship of deforestation with its direct and indirect determinants in the study area was analysed. Data were collected from the World Bank Data to analyse how the agricultural land, urban population growth, population growth and GDP per capita affect the forest areas in Malaysia from 1970 until 2019. The variables under direct factors are total agricultural land in Semenyih (AL), and the variables under indirect factors are urban population growth in Semnyih (UPG), total population growth in Selangor State (PG) and GDP per capita in Semenyih (GDPPC). Table 5.2 shows the hypothesis of the study. Miyamoto et al. (2014) demonstrated that direct and indirect factors caused deforestation, population growth and economic growth are the main factors that influence the forest cover in Peninsular Malaysia also identified that oil palm expansion as the direct factor of deforestation. Moreover, the effect of population, economic growth such as GDP and other economic factors are also the main causes of deforestation and carbon emission mainly contributed in Southeast Asia as well as Malaysia (Begum et al., 2020).

The drivers of deforestation used in this model are selected with the underlying factors which have the greater impact on deforestation such as economic factors, i.e. income and GDP, and physical factors, i.e. arable agriculture land, total primary forest areas, plantation forest areas and population (Indarto & Mutaqin, 2016). The

Fig. 5.5 Map of Selangor, Malaysia. (Source: Selangor Economic Planning Unit. Retrieved from https://www.selangor.gov.my/imageupload/Peta_Daerah_Negeri_Selangor-23Jul2006-064620.jpg)

environmental effect of deforestation was hypothesized by the Environmental Kuznets Curve (EKC) as in many previous studies (Tsiantikoudis et al., 2019; Mills Busa, 2013; Esmaeili & Nasrnia, 2014; Indarto & Mutaqin, 2016). The existence of EKC for deforestation is well-acknowledged, as well as mixed concepts of U-shaped and inverted U-shaped EKC. Most developed countries are still found

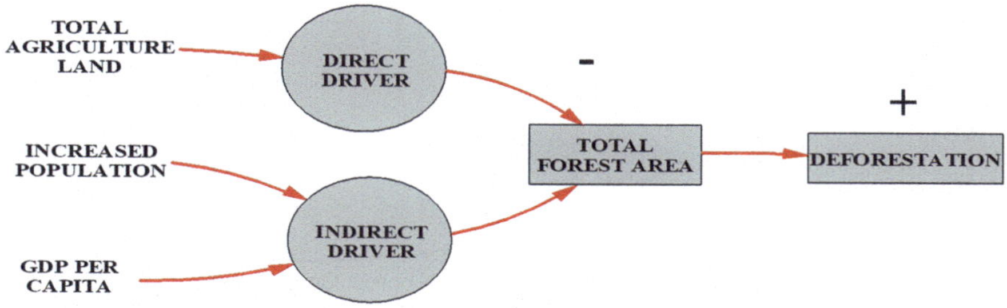

Fig. 5.6 Conceptual framework of the study

Table 5.2 Hypotheses of the study

Variables	Hypotheses
AL	H_{01}: There is a positive relationship between agricultural land (AL) and total forest area (FA) in Malaysia H_{A1}: There is no positive relationship between agricultural land (AL) and total forest area (FA) in Malaysia
GDPPC	H_{02}: There is a positive relationship between GDP per capita (GDPPC) and total forest area (FA) in Malaysia H_{A2}: There is no positive relationship between GDP per capita (GDPPC) and total forest area (FA) in Malaysia
UPG	H_{03}: There is a positive relationship between urban population growth (UPG) and total forest area (FA) in Malaysia H_{A3}: There is no positive relationship between urban population growth (UPG) and total forest area (FA) in Malaysia
PG	H_{04}: There is a positive relationship between population growth (PG) and total forest area (FA) in Malaysia H_{A4}: There is no positive relationship between population growth (PG) and total forest area (FA) in Malaysia

out with the U-shaped EKC; however, some empirical studies of deforestation effect on environment showed inverted U-shaped EKC in developing countries (Indarto & Mutaqin, 2016) with the increasing income and GDP. Deforestation is the main cause of reducing forest cover, and this is conceptualized through forest transition theory, which is the temporal changes of total forest areas or a change of forest cover trend over time (Miyamoto, 2020). The total forest cover is affected by many different direct and indirect factors. The transition of the forest is from moderate to high rate of deforestation (Fig. 5.7). The dynamic transition pattern of forest cover is shown with two trends or periods of forest decline and forest recovery stage. This is demonstrated through a reversed J-shaped curve (Perz, 2007). The recovery stage of forest cover could be sped up with afforestation or reforestation (Indarto & Mutaqin, 2016).

The impact of direct and indirect determinants of deforestation was analysed using ordinary least square (OLS) regression analysis, while the long- and short-run impacts were analysed with the autoregressive distributed lag (ARDL) model. Time series secondary data was used, and all of the variables were transformed into natural

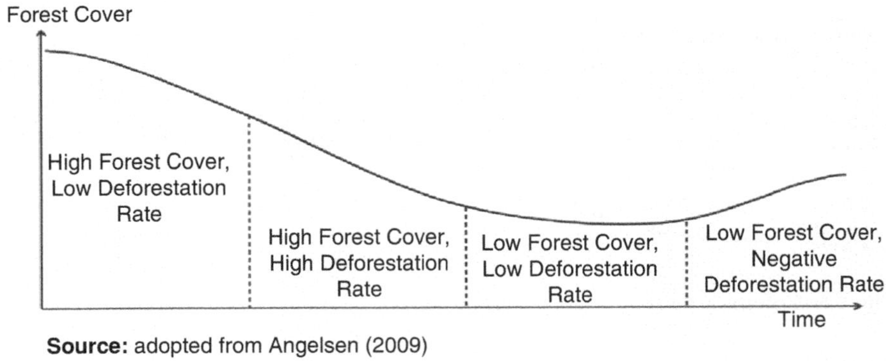

Fig. 5.7 Forest transition theory: temporal variation of forest cover. (Source: Indarto & Mutaqin, 2016)

logarithm form before analysis. OLS was used to estimate the multiple regression model as follows:

$$Y_i = \alpha + \beta_1 X_1 + \beta_2 X_2 + \beta_3 X_3 + \beta_4 X_4 + \varepsilon \tag{5.1}$$

where Y is forest area (FA), X_1 is agricultural land (AL), X_2 is urban population growth (UPG), X_3 is population growth (PG), X_4 is GDP per capita (GDPPC) and ε is the error term. The vector error correction model (VECM) approach was used to estimate the coefficients of the long-run relationship of the model. If the variables are co-integrated, long- and short-run causality can be detected by VECM. Before that, the short- and long-term effects of the variables were analysed using the following test:

1. ***Data stationary test or unit root test***
   Augmented Dickey–Fuller (ADF) method was used to test the stationarity of the variables in this model. Based on Baum (2013), the data was at the stationary level and did not have a unit root when the t statistics value was smaller than the critical value. The model cannot be used if the variables are not stationary because it will cause spurious regression in which R^2 is larger than Durbin Watson statistics. The model is a long-run model if the residuals are stationary which considered the variables are co-integrated. The regression for the unit root test (Phillips, 1987) is:

$$\Delta Y_t = \beta_1 + \beta_2 t + \delta Y_{t-1} + \sum_{i=1}^{m} \alpha_i Y_{t-i} + \varepsilon_t \tag{5.2}$$

   where ε_t = error term, β_1 = drift term, $\beta_2 t$ = time trend and Δ = differencing operator, respectively. The hypothesis for the unit root test is as follows:
   H_0: $\alpha = 0$ (series have a unit root and not stationary).
   H_1: $\alpha < 0$ (series does not have unit root test and stationary).

Reject the null hypothesis if the p value is less than 0.05 and the t statistics lies at the left of the critical value.

2. ***Lag length criteria***
   After the unit root test for each variable, the optimal length for this model was chosen to determine parameter estimates in VECM. The causality relationship was extremely affected by the lag length. The lowest value of information criteria, such as Akaike information criterion (AIC) and the Schwarz Bayesian criterion (SBIC), was chosen.
3. ***Co-integration test***
   Johansen co-integration test is used to study the long-term relationship between forest area, agricultural land, urban population growth, population growth and GDP per capita. As there was an integration of the same order for all the variables, determining the co-integration could be done. The equation for the co-integration test (Johansen, 1988) is as follows:

$$\Delta Y_t = \pi Y_{t-1} + \sum_{i=1}^{p-1} \pi_i \Delta Y_{t-i} + BX_t + \varepsilon_t \tag{5.3}$$

$$\pi = \sum_{i=1}^{p} A_i - I, \quad \pi_i = -\sum_{j=i+1}^{p} A_j \tag{5.4}$$

where Y_t = k-vector of non-stationary variables, X_t = d-vector of deterministic variables and ε_t = vector of white noises, respectively.

The hypothesis of the co-integration test is as follows:

H_0: $r = 0$ (no co-integration exists between dependent and independent variables).

H_1: $r \neq 0$ (co-integration exists between dependent and independent variables). The trace test was preferred although the trace and maximum eigenvalue test generated different results.

5.3 Research Findings: The Impacts of Direct and Indirect Factors on Deforestation

The relationship between direct determinants, i.e. agricultural land (AL), and indirect determinants, i.e. urban population growth (UPG), population growth (PG) and GDP per capita (GDPPC), with the indicator variable of total forest area was analysed using multiple regression analysis. The regression results are shown in Table 5.3. AL has a negative, significant relationship with the total forest area in which the increasing number of agricultural lands causes the increasing deforestation (Zulkofli et al., 2018). Both UPG and PG also have a negative, significant relationship with the total forest area with a 1% and 5% significance level. This implies that population growth induces the loss of total forest areas and inherently increases deforestation (Miyamoto, 2020). The increase in deforestation rate is also caused by population

Table 5.3 Multiple regression results of the AL, UPG, PG, GDPPC and total forest area

Variable	Coefficient	Std. error	t-Statistic	Prob.
Dependent variable: FA				
C	13.705	4.653	−2.945	0.005
LNAL	−2.122	0.354	5.989	0.000***
LNUPG	−0.605	0.236	2.563	0.014**
LNPG	−0.976	0.297	−3.281	0.002***
LNGDPPC	−0.323	0.092	−3.494	0.001***
Adjusted R-squared	0.716			
Durbin Watson stat	0.735			

Notes: *Significant level of 0.10, **significant level of 0.05, ***significant level of 0.01

growth and increasing demand for food, fuel, wood, timber and other forest products (Angelsen & Kaimowitz, 1999). Meanwhile, the coefficient value of −0.323 implies a negative relationship between GDPPC and total forest area. The 1% increase in GDPPC leads to the decrease of forest areas by 0.32%. In agreement with this, Crespo Cuaresma et al. (2017) found that forest cover is negatively correlated with GDPPC.

Based on results in Table 5.3, the significant relations were found in AL, UPG, PG and GDPPC with total forest area of 1%, 5%, 1% and 1%, respectively, and the regression model could be constructed as follows:

$$\begin{aligned}\text{LNFA} = 13.705 - 2.122\ \text{LNAL} - 0.605\ \text{LNUPG} - 0.976\ \text{LNPG}\\ - 0.323\ \text{LNGDPPC} + \varepsilon\end{aligned} \tag{5.5}$$

where FA is a total forest area, UPG is urban population growth, PG is total population growth and GDPPC is GDP per capita, respectively.

5.3.1 Short- and Long-Run Relationship of Direct and Indirect Determinants of Deforestation

The short- and long-run relationship of the direct determinant (AL) and indirect determinant (UPG, PG and GDP) with total forest area was analysed using the ARDL model. The unit root test shows that at level differences, all variables are not stationary and are integrated at the same order in the first difference as illustrated in Table 5.4.

When the variables are tested for the unit root effect, the ARDL model with an appropriate lag difference is constructed. If all the variables are integrated in the same order in the ADF test, the analysis continues to identify the co-integration level using the Johansen–Juselius co-integration test which is the vector error correction model (VECM) approach. The optimum lag of this model is lag 3. The standard ARDL with optimum lag is as follows:

Table 5.4 Unit root test result

Variable	Augmented Dickey–Fuller (ADF)			
	Level		First difference	
	Constant without trend	Constant with trend	Constant without trend	Constant with trend
LNFA	−1.371	−1.949	−6.500***	−6.423***
LNAL	−1.026	−2.183	−2.219	−4.411***
LNUPG	1.550	−0.679	−4.962***	−5.333***
LNPG	2.020	−0.766	−6.514***	−8.431***
LNGDPPC	−1.989	−2.434	−6.287***	−6.671***

Note: *** and ** denote significance at 1% and 5% level, respectively

Table 5.5 Long-run causality test result

	Coefficient	Std. error	*t*-Statistic	Prob.
C (1)	**– 0.018****	0.008	– 2.314	0.029
C (2)	**– 0.491*****	0.169	– 2.903	0.007
C (7)	**– 2.703****	1.239	– 2.181	0.038

Note: *** and ** denote significance at 1% and 5% level, respectively

$$\begin{aligned} & \mathrm{D(LNFA)C} + \mathrm{D(LNFA(-1))} + \mathrm{D(LNFA(-2))} + \mathrm{D(LNFA(-3))} \\ & + \mathrm{D(LNAL(-1)} + \mathrm{D(LNAL(-2))} + \mathrm{D(LNAL(-3))} \\ & + \mathrm{D(LNUPG(-1))} + \mathrm{D(LNUPG(-2))} + \mathrm{D(LNUPG(-3))} \\ & + \mathrm{D(LNPG(-1))} + \mathrm{D(LNPG(-2))} + \mathrm{D(LNPG(-3))} \\ & + \mathrm{D(LNGDPPC(-1))} + \mathrm{D(LNGDPPC(-2))} + \mathrm{D(LNGDPPC(-3))} \\ & + \mathrm{(LNFA(-1)} + \mathrm{(LNAL(-1)} - \mathrm{LNUPG(-1)} + \mathrm{LNPG(-1)} \\ & - \mathrm{LNGDPPC(-1)} + \varepsilon \end{aligned} \tag{5.6}$$

The long-run association between the dependent variable and independent variable is tested using the co-integration test. The null hypothesis is rejected if there is no co-integration and the value of the trace statistics is larger than the 5% level critical value. The results in Table 5.5 show that trace statistics are statistically significant to reject the null hypothesis of $r = 0$ at a 5% significance level. There is one co-integration equation. The significant associations were found on C (1): LNFA (– 1), C (2): LNAL (– 1) and C (7): LNAL (– 3); therefore, a relationship does exist between forest area and agricultural land.

In this study, the *F*-statistics values were also evaluated with the Pesaran critical value. They were unrestricted intercept, and no trend models were supposed to show the *F*-statistics value of less than the lower bond value of 3.79 or greater than the upper bound value of 4.85. The results show that all coefficient values of the variables are less than the lower bound value. Therefore, the model has no short-run association between independent variables of AL, UPG, PG and GDPPC to the forest area. In other words, all the direct and indirect determinants show only long-run effects on the total forest areas (Table 5.6).

Table 5.6 Short-run causality test result

Normalized restriction (=0)	*F*-statistics probability	Chi-square probability
C (5) = C (6) = C (7)	0.084	0.065
C (8) = C (9) = C (10)	0.955	0.955
C (11) = C (12) = C (13)	0.996	0.996
C (14) = C (15) = C (16)	0.832	0.831

Table 5.7 Model validity test result

Diagnosis tests	Results	Decisions
Heteroskedasticity	*F*-statistics = 2.581 *P*-value = 0.116	*P*-value (0.116) > 0.01 H_0 accepted Thus, residuals have no heteroscedasticity
Serial correlation	*F*-statistics = 0.910 *P*-value = 0.416	*P*-value (0.416) > 0.01 H_0 accepted Thus, residuals have no serial correlations

Source: Authors' own research

5.3.2 Residual Diagnosis and Model Validation

Based on the LM (*Lagrange Multiplier*) serial autocorrelation test above, there is a probability of chi-square equal to 0.220. It shows that the value of chi-square is more than the significant level which is 0.05. Autocorrelation can also be detected by Durbin Watson which consists of upper and lower bounds, dU and dL. If d is more than dU, then this model will consider no autocorrelation. In this case, dU is 1.528 at a 1% significant level and 1.72 at a 5% significant level. Hence, H_0 was not rejected, showing that the model did not suffer from autocorrelation. Homoscedasticity is an equal variance of the disturbance term, while heteroscedasticity is a different variance of the error term. The result shows that there is a probability that chi-square is equal to 0.111 and it is more than the significant level which is 0.05. Hence, H_0 was not rejected, showing that this model did not suffer from a heteroscedasticity problem (Table 5.7).

5.4 Discussion

Forests cover over one-third of the globe, yet they are rapidly disappearing. A forest is often associated with natural, man-made and environmental stressors. The ability of the forest to resist and adapt to these stressors is an important indicator of forest health. At certain places, people depend on forest resources to lift themselves out of poverty and reduce their vulnerability. Concerning the continuous threat of the loss of forest areas across the globe and the significant challenges to protect forests and their ecosystem services, related stakeholders, i.e. government, private sector and community, must have proper forest ecosystem management to ensure forest sustainability and development.

5.4.1 Government

Forest management began in the early 1900s in Malaysia. All Malaysian states have the authority to decide on forest resource administration, management, usage and allocation. However, each state in Malaysia has the authority to set its forestry legislation, where Malaysia has two major forest policies. The two main laws governing forestry activities in Malaysia are the National Forestry Act of 1984 and the National Forestry Policy of 1978. These policies are intended to safeguard forest management areas against unlawful harvesting, unauthorized settlement and other prohibited activities.

The majority of forest land areas are owned by the government; thus, it has the right to protect and conserve the land by applying certain policies for the benefit of the forest and its surroundings. For instance, the government should identify which forest land cannot be transformed into agricultural land without permission. Among the roles that the government can play here are gathering information and data, disseminating technical information, implementing research and organizing institutions involved. The Energy and Natural Resources Ministry (KeTSa), with cooperation from the Forestry Department of Peninsular Malaysia (JPSM) and Forest Research Institute of Malaysia (FRIM), has organized "100 Million Tree Planting" campaign to raise awareness on the importance of green areas and forests to ensure environmental sustainability, well-protected biodiversity and the good quality of life within the community. This is in line with the slogan "Greening Malaysia: Our Trees, Our Life". It is targeted that 100 million trees will be planted in 5 years starting from 2021 until 2025. Furthermore, this is also in line with JPSM's mission to manage and develop forest resources sustainably and optimize their contributions to socio-economic development. The campaign will protect the country's biological diversity as a valuable national treasure and improve the quality of the river.

5.4.2 Private Sector

The collaboration and financial contribution from the private sector can safeguard the future of the forests. Hence, it is significant to propose a framework that focuses on the investment from private companies, as well as new incentives to encourage public–private partnerships on forests. For instance, forest conservation financing is a good move by encouraging private companies and local communities to work together to protect the forests. In this sense, the payment ecosystem services (PES) scheme is a unique mechanism for environmental protection. When a user of an ecosystem service makes a direct or indirect payment to the provider of that service, this is referred to as payment for ecosystem service. This will encourage the natural ecosystems to be persevered and promote ecological benefits that do not harm the users of natural resources. In the private sector, the recipients of ecosystem services are willing to pay a price that is expected to be less than their welfare gain because of the services. The providers of ecosystem services should be willing to accept a payment that exceeds the cost of providing the services. The PES demonstrates that in a competitive market with no transaction costs and clear property rights, direct negotiation between private parties can result in efficient consequences (Coase, 1960). Funding from private companies (buyers) for carbon sequestration, for example, in the case of Bogor Agricultural University (IPB) as a provider through the PES scheme at Gunung Walat University Forest (GWUF), Indonesia, was a success in transforming bare lands into forested lands (Tiryana et al., 2014). Furthermore, the enrichment planting activities targeted at degraded forest areas in GWUF, whereas the funding was raised from private sector participants, i.e. TOSO, Conoco Phillip and NYK with a voluntary scheme within a contract period of one to five years. At the same time, the companies have effectively invested in corporate social responsibility (CSR) funding for educational and environmental purposes, enhanced green company image branding and participated in reducing CO_2 emissions.

5.4.3 Community

Community involvement is crucial to any conservation efforts. Participation in forest conservation is regularly associated with community forestry which refers to forest management or co-management by communities who live near the forest. Conservation without community's participation is ruined in certain cases. Community forestry (CF) is commonly associated with South and Southeast Asia. It can also be found in other parts of the world (Wily, 1997). The CF is adapted through partnerships and coordination between communities and forest landowners to stimulate forest stewardship and economic development. It should be managed by the local communities to highlight the collaborative and participatory management of local needs and knowledge. The CF is also effective at lessening resource degradation and supporting biodiversity conservation (Adhikari et al., 2004). An effective CF management generates sustainable production and harvesting of forest resources. Conservation without community's participation is ruined in certain cases.

The natural attractions developed are expected to realize the sustainable management of natural resources through the development of ecotourism. Ecotourism influences social, environmental and economic aspects and contributes to the conservation of biodiversity. Ioki et al. (2019) and Hassin et al. (2020) have researched ecotourism resources through community based tourism (CBT) in Crocker Range Park, Sabah, and Gelam Forest, Kelantan, revealing that ecotourism appears to play an important role in protecting the natural environment. It measures to effectively control logging and encourage local communities to take part in forest sustainability campaigns. Nonetheless, the choice to implement CF management is based on the community's acceptance and commitment to resource management. Therefore, the local communities should alert themselves about the connection with ecosystems to encounter the ecological, social and economic changes from time to time.

5.5 Conclusion

Deforestation is the loss of total forest areas due to various direct and indirect determinants. Peninsular Malaysia is currently suffering a rapid loss of forest areas due to the increase of agricultural land (direct determinant), urban population growth, total population and GDP per capita (indirect determinants). The analysis of the impact of these direct and indirect determinants of deforestation in Peninsular Malaysia pointed out that there is a long-run impact of these determinants on total forest area and they accelerated the rate of deforestation. The long-run impact of all drivers, both direct and indirect drivers such as agriculture land (AL), urban population growth (UPG), total population growth (PG) and GDP per capita (GDPPC), to the forest area in the study area highlights the managing of forest area in long-term planning to maintain the urban forest in Peninsular Malaysia. In other words, all the direct and indirect determinants show only long-run effects on the total forest areas. The long-run correlation of the maintenance of urban forest, primary forest (natural forest) and plantation forest is a crucial issue to be emphasized and managed with effective policies and sustainable forest management practices, for example, silviculture treatment and afforestation or reforestation in the urban forest conservation.

At the same time, Sustainable Development Goals (SDGs) offer an inclusive mechanism by tackling social, economic and environmental concerns and dimensions, with attention given to sustainability and human rights. Forest management is included in the SDGs, for example, sustainable forest management, biodiversity protection, water sources, flood control measures, sustainable tourism and lifestyle. For instance, SDG15 (Life on Land) targets to "protect, restore, and promote sustainable use of terrestrial ecosystems, manage forests sustainably, combat desertification, and halt and reverse land degradation and biodiversity loss". Human-caused deforestation and desertification pose great challenges to sustainable development and have impacted the livelihood of millions of people. Forests are critical to the survival of life on Earth and play an important role in the fight towards climate change. Furthermore, investment in land restoration is crucial for livelihood and lessening the economic risks.

References

Adhikari, B., Di Falco, S., & Lovett, J. C. (2004). Household characteristics and forest dependency: Evidence from common property forest management in Nepal. *Ecological Economics, 48*, 245–257. https://doi.org/10.1016/j.ecolecon.2003.08.008

Aisyah, A., Shahrul, A. B., Zulfahmie, M. Z. M., et al. (2015). Deforestation analysis in Selangor, Malaysia between 1989 and 2011. *Journal of Tropical Forest Science, 27*(1), 3–12.

Angelsen, A., & Kaimowitz, D. (1999). Rethinking the causes of deforestation: Lessons from economic models. *The World Bank Research Observer, 14*, 73–98. https://doi.org/10.1093/wbro/14.1.73

Assunção, J., Gandour, C., Rocha, R., & Rocha, R. (2019). The effect of rural credit on deforestation: Evidence from the Brazilian Amazon. *The Economic Journal, 130*, 290–330. https://doi.org/10.1093/ej/uez060

Austin, K. G., Schwantes, A., Gu, Y., & Kasibhatla, P. S. (2019). What causes deforestation in Indonesia? *Environmental Research Letters, 14*, 024007. https://doi.org/10.1088/1748-9326/aaf6db

Baum, C. F. (2013). *VAR, SVAR and VECM models. EC 823: Applied econometrics*. Boston College.

Begum, R. A., Raihan, A., & Said, M. N. (2020). Dynamic impacts of economic growth and forested area on carbon dioxide emissions in Malaysia. *Sustainability, 12*, 9375. https://doi.org/10.3390/su12229375

Bhatia, N., & Cumming, G. S. (2020). Deforestation and economic growth trends on oceanic islands highlight the need for meso-scale analysis and improved mid-range theory in conservation. *Ecology and Society, 25*, 250310. https://doi.org/10.5751/es-11713-250310

Bologna, M., & Aquino, G. (2020). Deforestation and world population sustainability: A quantitative analysis. *Scientific Reports, 10*, 7631. https://doi.org/10.1038/s41598-020-63657-6

Chazdon, R. L. (2003). Tropical forest recovery: Legacies of human impact and natural disturbances. Perspectives in plant ecology. *Evolution and Systematics, 6*, 51–71. https://doi.org/10.1078/1433-8319-00042

Christian, A. K. (2017). Forest transitions: A new conceptual scheme. *Geographica Helvetica, 72*, 465–474. https://doi.org/10.5194/gh-72-465-2017

Coase, R. H. (1960). The problem of social cost. *The Journal of Law and Economics, 3*, 63–94. https://doi.org/10.1007/978-1-349-12486-2_4

Crespo Cuaresma, J., Danylo, O., Fritz, S., et al. (2017). Economic development and forest cover: Evidence from satellite data. *Scientific Reports, 7*, 40678. https://doi.org/10.1038/srep40678

Department of Statistics Malaysia. (2021). *Gross Domestic Product (GDP) by State 2021*. Extracted and downloaded [July 2021] from https://www.dosm.gov.my/v1/index.php?r=column/cthemeByCat&cat=491&bul_id=YnhhZ2g5QlpZWG9RcVNwTGhLaHE4UT09&menu_id=TE5CRUZCblh4ZTZMODZIbmk2aWRRQT09

DOSM. (2020). Department of Statistics Malaysia. Retrieved from https://www.dosm.gov.my

Esmaeili, A., & Nasrnia, F. (2014). Deforestation and the environmental Kuznets curve in Iran. *Small-Scale Forestry, 13*, 397–406. https://doi.org/10.1007/s11842-014-9261-y

FAO. (2016). *Global forest resources assessment 2015: How are the world's forests changing?* (2nd ed.). Food and Agriculture Organization of the United Nations.

Ferronato, N., & Torretta, V. (2019). Waste mismanagement in developing countries: A review of global issues. *International Journal of Environmental Research and Public Health, 16*, 1060. https://doi.org/10.3390/ijerph16061060

GFW, Global Forest Watch. (2021). *Assessing trends in tree cover loss over 20 years of data*. Retrieved September 30, 2021, from https://www.globalforestwatch.org/blog/data-and-research/tree-cover-loss-satellite-data-trend-analysis/

Giardina, C. P., Sanford, R. L., Døckersmith, I. C., & Jaramillo, V. J. (2000). The effects of slash burning on ecosystem nutrients during the land preparation phase of shifting cultivation. *Plant and Soil, 220*(1), 247–260. https://doi.org/10.1023/A:1004741125636

Global Forest Watch Malaysia. (2021). *Vizzuality Malaysia deforestation rates and statistics: GFW*. Global Forest Watch. Retrieved September 30, 2021, from https://www.globalforestwatch.org/dashboards/country/MYS?treeLossLocated=eyJwYWdlIjozfQ%3D%3D%26treeLossTsc

Hancock, L. (2021). *Deforestation and forest degradation*. World Wildlife Fund. Retrieved September 30, 2021, from https://www.worldwildlife.org/threats/deforestation-and-forest-degradation

Hansen, M. C., Potapov, P. V., & Moore, R. (2015). *Tree cover loss and gain area*. University of Maryland, Google, USGS, and NASA. Retrieved January 16, 2015, from www.globalforestwatch.org

Hassin, N. H., Koshy, N., Hambali, K., & Kumaran, J. V. (2020). Local communities willingness to pay for conservation of ecotourism resources at Gelam Forest, Kelantan, Malaysia. *IOP Conference Series: Earth and Environmental Science, 549*, 012090. https://doi.org/10.1088/1755-1315/549/1/012090

Hattori, D., Kenzo, T., Shirahama, T., et al. (2019). Degradation of soil nutrients and slow recovery of biomass FOLLOWING shifting cultivation in the heath forests of Sarawak, Malaysia. *Forest Ecology and Management, 432*, 467–477. https://doi.org/10.1016/j.foreco.2018.09.051

Hepp, C. M., de Neergaard, A., & Bruun, T. B. (2018). Short-term fallow in extensive upland shifting cultivation systems of Northern Lao PDR: Its role in soil fertility restoration. *Land Degradation & Development, 29*, 2911–2919. https://doi.org/10.1002/ldr.3032

Indarto, J., & Mutaqin, D. J. (2016). An overview of theoretical and empirical studies on deforestation. *Journal of International Development and Cooperation., 22*(1 & 2), 107–120.

Ioki, K., Din, N. M., Ludwig, R., et al. (2019). Supporting forest conservation through community-based land use planning and participatory GIS—Lessons from Crocker Range Park, Malaysian Borneo. *Journal for Nature Conservation, 52*, 125740. https://doi.org/10.1016/j.jnc.2019.125740

Ishizuka, S., Sakurai, K., Sabang, J., et al. (2000). Soil characteristics of an abandoned shifting cultivation land in Sarawak, Malaysia. *Tropics, 10*, 251–263. https://doi.org/10.3759/tropics.10.251

Jellason, N. P., Robinson, E. J., Chapman, A. S., et al. (2021). A systematic review of drivers and constraints on agricultural expansion in Sub-Saharan Africa. *Land, 10*, 332. https://doi.org/10.3390/land10030332

Johansen, S. (1988). Statistical analysis of cointegrating vectors. *Journal of Economic Dynamics and Control, 12*, 231–254. https://doi.org/10.1016/0165-1889(88)90041-3

Kleinman, P. (1995). The ecological sustainability of slash-and-burn agriculture. *Agriculture, Ecosystems & Environment, 52*, 235–249. https://doi.org/10.1016/0167-8809(94)00531-i

Lamb, D. (2010). *Regreening the bare hills: Tropical forest restoration in the Asia-Pacific region* (Vol. 8). Springer.

Malmer, A., Johansson, E., & Kluge, M. (1998). Natural rehabilitation of disturbed tropical rainforest soils in Sabah, Malaysia. In *Soils of tropical forest ecosystems* (pp. 199–204). Springer. https://doi.org/10.1007/978-3-662-03649-5_23

Millennium Ecosystem Assessment Panel. (2005). *Millennium Ecosystem Assessment Panel (MEA)*. Island Press.

Mills Busa, J. H. (2013). Dynamite in the EKC tunnel? Inconsistencies in resource stock analysis under the environmental Kuznets curve hypothesis. *Ecological Economics, 94*, 116–126. https://doi.org/10.1016/j.ecolecon.2013.07.015

Miyamoto, M. (2020). Poverty reduction saves forests sustainably: Lessons for deforestation policies. *World Development, 127*, 104746. https://doi.org/10.1016/j.worlddev.2019.104746

Miyamoto, M., Mohd Parid, M., Noor Aini, Z., & Michinaka, T. (2014). Proximate and underlying causes of forest cover change in Peninsular Malaysia. *Forest Policy and Economics, 44*, 18–25. https://doi.org/10.1016/j.forpol.2014.05.007

Omar, H., & Misman, M. A. (2018). Time-series maps of aboveground biomass in dipterocarps forests of Malaysia from PALSAR and PALSAR-2 polarimetric data. *Carbon Balance and Management, 13*(1), 1–9. https://doi.org/10.1186/s13021-018-0108-2

Perz, S. G. (2007). Grand theory and context-specificity in the study of forest dynamics: Forest transition theory and other directions. *The Professional Geographer, 59*(1), 105–114.

Phillips, P. C. B. (1987). Time series regression with a unit root. *Econometrica, 55*(2), 277–301. https://doi.org/10.2307/1913237

Pugh, T. A. M., Lindeskog, M., Smith, B., Poulter, B., Arneth, A., Haverd, V., & Calle, L. (2019). Role of forest regrowth in global carbon sink dynamics. *Proceedings of the National Academy of Sciences of the United States of America, 116*(10), 4382–4387. https://doi.org/10.1073/pnas.1810512116

Shevade, V. S., & Loboda, T. V. (2019). Oil palm plantations in Peninsular Malaysia: Determinants and constraints on expansion. *PLoS One, 14*(2), e0210628. https://doi.org/10.1371/journal.pone.0210628

Shevade, V. S., Potapov, P. V., Harris, N. L., & Loboda, T. V. (2017). Expansion of industrial plantations continues to threaten Malayan tiger habitat. *Remote Sens, 9*(7), 747.

Tanaka, S., Kendawang, J. J., Yoshida, N., et al. (2005). Effects of shifting cultivation on soil ecosystems in Sarawak, Malaysia IV. Chemical properties of the soils and runoff water at Niah and Bakam experimental sites. *Soil Science and Plant Nutrition, 51*, 525–533. https://doi.org/10.1111/j.1747-0765.2005.tb00061.x

Tiryana, T., Kosmaryandi, N., Sukendro, A., et al. (2014). Voluntary carbon project in Gunung Walat University Forest. In *The 3rd Annual Meeting of the Low Carbon Asia Research Network (LoCARNet), Bogor, Indonesia, 24–26 November 2014*. Retrieved from https://lcsrnet.org/pdf/lcs_rnet_presentations/6th/P2.C-3_Tiryana.pdf

Tsiantikoudis, S., Zafeiriou, E., Kyriakopoulos, G., et al. (2019). Revising the environmental Kuznets curve for deforestation: An empirical study for Bulgaria. *Sustainability, 11*, 4364. https://doi.org/10.3390/su11164364

United Nations. (2021). *SDG indicators: Metadata repository*. In: United Nations, Department of Economic and Social Affairs, Statistics Division [online]. New York, USA. https://unstats.un.org/sdgs/metadata

Villa, P. M., Martins, S. V., de Oliveira Neto, S. N., et al. (2018). Intensification of shifting cultivation reduces forest resilience in the Northern Amazon. *Forest Ecology and Management, 430*, 312–320. https://doi.org/10.1016/j.foreco.2018.08.014

Wily, .L (1997). *Villagers as forest managers and governments "learning to let go": The case of Duru-Haitemba and Mgori forests in Tanzania* (Forest participation series no. 9). International Institute for Environment and Development

WWF. (2021). Deforestation front: Drives and responses in a changing world (summary). In P. Pacheco, K. Mo, N. Dudley, A. Shapiro, N. Aguilar-Amuchastegui, P. Y. Ling, C. Anderson, & A. Marx (Eds.), *Deforestation fronts: Drivers and responses in a changing world*. WWF.

Ying, H. S., Wasli, M. E., & Perumal, M. (2018). Soil characteristics under intensified shifting cultivation for upland rice cultivation in upland Sabal, Sarawak, Malaysia. *BIOTROPIA-The Southeast Asian Journal of Tropical Biology, 25*(1), 72–83. https://doi.org/10.11598/btb.2018.25.1.799

Young, C., SACCESS & JKOASM. (2015). Deforestation drivers and human rights in Malaysia. In *International workshop on deforestation drivers and the rights of forest people, Palangkaraya, Indonesia, March 9–14, 2015*. Human Rights Documents online. https://doi.org/10.1163/2210-7975_hrd-5540-2014003

Zulkofli, A. Z., Nasiruddin, K., & Ismail, F. R. (2018). Relationship between forest area with population, electricity and agriculture in Malaysia. *International Journal of Accounting & Business Management, 6*(1), 113–124. https://doi.org/10.24924/ijabm/2018.04/v6.iss1/113.124

A Trade-Off Analysis of Sustainable Landscape Planning: A Case Study of Sintang Regency (Heart of Borneo), Kalimantan

6

Sopian Hidayat, Rina Wulandari, Noor Aini Zakaria, and Christy Desta Pratama

Abstract

Forest in Indonesia is among one of the richest biodiversity in the world. However, this forest is currently under threat due to several factors such as illegal logging, forest fire, and forest conversion. Considering the ecological and economic importance of this forest, the local government is committed to its protection and sustainable use under the umbrella of "Green Regency" and has been working with several organizations to develop long-term landscape planning. This research focuses on Sintang, West Kalimantan, that has a vast tropical forest with high biodiversity of flora and fauna. In 2016, the Sintang Regency has declared its commitment as a sustainable regency and committed to protecting the forest within the framework of the Sustainable Development Goals (SDG). The research aims are to analyze land use and land cover (LULC) changes from 2006 to 2016 and forecast the land use and land cover on business as usual (BAU) and green scenario. The analysis used Land Change Modeler (LCM) module in TerrSet software to project the land cover. Based on the analysis of land cover change from 2006 to 2016, the secondary forest area experienced the largest decrease (−87,680 ha), while the plantation area had the largest increase (87,540 ha). The BAU scenario projected that plantations would grow to 253,571 ha in 2030 while only 180,300 ha under the Sintang Lestari scenario (green scenario). The results of the Sintang Lestari scenario showed that

S. Hidayat (✉) · C. D. Pratama
Conservation Strategy Fund (CSF) Indonesia, Jakarta, Indonesia

R. Wulandari
Independent Researcher, Bogor Utara, Jawa Barat, Indonesia

N. A. Zakaria
Malaysian Environmental Economics Association (MEEA), c/o Institute of Forestry and Forest Products (INTROP), Universiti Putra Malaysia (UPM), Serdang, Selangor, Malaysia

Z. Samdin et al. (eds.), *Tropical Forest Ecosystem Services in Improving Livelihoods For Local Communities*, https://doi.org/10.1007/978-981-19-3342-4_6

significant plantation development can still be achieved while protecting the remaining forest cover. However, limiting the expansion of plantation areas is only possible if current spatial plans are monitored and non-compliance with spatial planning is strictly enforced. The findings of the analysis can serve as a strong basis for future land development in Sintang Regency.

Keywords

Sustainable landscape · Land use · Land cover · Sustainable Development Goals (SDG) · Indonesia

6.1 Introduction

Forest plays a major role in supporting human life. Forest provides its services through the source of food and energy, reserving water supply, maintaining soil fertility, preventing humans from natural disasters such as floods and land erosion control, education and research, natural tourism, and others (García-Nieto et al., 2013; Nowak et al., 2015; Brockerhoff et al., 2017; Aznar-Sánchez et al., 2018). García-Nieto et al. (2013) and Brockerhoff et al. (2017) highlighted that forest diversity plays an important factor in sustaining human life through ecosystem function and the provision of ecosystem services. Protecting the forest means preserving the natural habitats and their ecosystem including sustaining the human-life balance (Aznar-Sánchez et al., 2018).

In Indonesia, as in many other parts of the world, population growth, economic growth, and climate change issues turn out to increase pressures on natural resources. These challenges need to be addressed simultaneously. Studies have proven the relationship between economic growth and CO_2 emissions (Fankhauser & Tol, 2005; Hwang & Yoo, 2012; Shahbaz et al., 2013; Knight & Schor, 2014). Higher growth rates result in higher emissions, which contribute to climate change. Climate change has a direct impact on the economy and caused global warming and reduced future welfare (Fankhauser & Tol, 2005). Future development should be consistent with the implementation of the green growth plan while simultaneously focusing on economic growth with low carbon emissions and, finally, achieving the Sustainable Development Goals (Noorbakhsh & Ranjan, 1999; Knight & Schor, 2014). Sustainable development is a development that meets the needs of the present without compromising future generations' ability to meet their own needs (Pearce et al., 1994), and natural capital owned by regions must be preserved in the long run (Noorbakhsh & Ranjan, 1999).

Indonesia forest stands are under severe threat due to deforestation and forest degradation which have occurred in both mineral soil and peatland. Deforestation has contributed to CO_2 emissions as a result of forest conversion to other land uses particularly oil palm plantation expansion which creates an additional source of greenhouse gas emissions (Schebek et al., 2018). Emission from mineral lands mostly comes from forest encroachment, forest conversion to other land use, and forest fires (Malahayati & Masui, 2019; Afriyanti et al., 2019). Emission from peatlands originates in forest conversion, degradation, peat decomposition, peat fires, and mangrove conversion. Margono et al. (2014) highlighted that Indonesia forest cover loss between 2000 and 2012 was about 16 million ha of which 38% of it occurred within the Indonesian primary forest. From that period, the average forest cover loss was increased by 47,600 ha/year. Kalimantan Island is the second-highest forest loss after Sumatra Island. Most of the primary forest loss occurred in the lowland areas. Studies indicate that agricultural area expansion is the main driving factor causing deforestation in Kalimantan (Gunarso et al., 2013; Laurance et al., 2014; Margono et al., 2014; Abram et al., 2017). The remaining tropical forest in Kalimantan is under threat caused by the growing agricultural demand in the region.

6.1.1 Sintang and Sustainable Regency (Sintang Lestari)

Sintang is one of the regencies in West Kalimantan that has a vast tropical forest with high biodiversity of flora and fauna. In 2019, it has a total population of 403,000 people and a population density of 19 people/km^2 (Central Bureau of Statistic Indonesia, 2020). Of the total area (21,638 km^2), 59% of the regency area is designated as a forest area, and it also has a national park located in the southern part of the regency (Bukit Baka Bukit Raya National Park). The total forest area is 1,286,944 ha which consists of limited production forest (28%), protected forest (21%), production forest (6%), conservation area (3%), and convertible production forest (1%). Among the forested area, 396,566 ha is categorized as primary dryland forest and mostly located in the protected forest, limited production forest, and conservation areas. Despite its large designated forest area, the Sintang Regency experiencing the highest rates of deforestation, forest degradation, and forest fires. Deforestation in Sintang is primarily caused by high demand for land-based products (agriculture, plantation, and mining), encroachment, and other economic activities. Most people rely on agriculture as a source of income, and the agricultural sector plays an important role in the Sintang economy with high job opportunities. However, Sintang Regency faces serious development challenges, including the poverty rate of its population, which is currently among the highest in West Kalimantan, economic growth that is still below target, inadequate and inequitable basic infrastructure provision, and other issues. At the same time, pressure on natural resources is increasing. More than a hundred thousand indigenous people who live in the forest rely on it for a living, and most of them rely on the environmental services it provides, such as clean water and high air quality.

Since declaring its commitment as a Sustainable Regency in 2016, the local government has continued to improve itself and strive to provide greater welfare for its people while still paying attention to the sustainability of natural resources, particularly forests and land, and environmental conditions. This commitment certainly requires synergy between all elements of stakeholders in the Sintang District, including the district government and civil society. In addition, Sintang Regency has committed its district to move sustainably in the landscape and development planning. In light of the ecological and economic importance of the forest, the local government is committed to its protection and sustainable use under the framework of Sustainable Development Goals (SDG). With the commitment, the Sintang Government had conducted a multi-stakeholder and institutional design towards the sustainable regency and green growth pathways. This process accumulated stakeholders' voices in formulating a shared vision of the future landscape, with the forested area remaining constant or even expanding due to no room for deforestation. In addition, the government commitment to the green pathway needs to be assessed as the regency relies on natural and land-based resources for attracting economic growth.

In general, the purpose of this research is to conduct a trade-off analysis of sustainable landscape planning of the Sintang Regency. The research has two goals: (1) to analyze land use and land cover (LULC) changes from 2006 to 2016 (past 15 years) and (2) to project the land use and land cover on business as usual (BAU) and green scenario from 2016 to 2030. The outcomes of this research will be utilized to help the government make decisions and comprehend past land use changes. Furthermore, this will serve as a solid foundation for future land use projects aimed at achieving sustainable regency.

6.2 Data and Methodology

6.2.1 Study Area

The study covers an area of Sintang Regency, West Kalimantan Province. Sintang Regency is one of 14 regencies in West Kalimantan Province with a high forest cover. Sintang Regency is the third largest district in West Kalimantan Province, after Ketapang Regency (31,240 km^2) and Kapuas Hulu Regency (29,842 km^2). Sintang Regency has a total area of 21,638 km^2 and is divided into 14 sub-districts. The regency is endowed with an abundance of natural resources and forest ecosystems, which serve as the economic foundation for both the government and the local community. The topographic condition of the Regency is mostly hilly (62.74%) with elevation ranging from 8 to 2040 m above sea level (Fig. 6.1).

Sintang has two major rivers, Kapuas and Melawi, which were once the primary modes of community transportation. Currently, Sintang Regency's total forest area is 59% of its total area, accounting for 1.3% of Indonesia's total forest area (Central Bureau of Statistic Indonesia, 2020). Sintang Regency is committed to achieving sustainable development by preserving the remaining forest area.

6.2.2 Data

Data is gathered from both primary and secondary sources. Primary data were obtained through direct interviews with government stakeholders of the Sintang Regency. The study used secondary data, both spatial and non-spatial data. Spatial

Fig. 6.1 Sintang Regency. (Source: Indonesia Geospatial Agency (2006))

data includes administration boundary, series of the land cover map, and forest status. Administration data and base map derived from Indonesia Geospatial Agency (BIG), while data related to land cover and forest status derived from the Ministry of Environment and Forestry (MoEF). Land cover data from 2006 to 2016 were used in this study to analyze the trend of land use and land cover changes (LULCC), which was then used as the basis for the land use projection model for the year 2030 (Table 6.1).

Table 6.1 Data description

Data	Description	Source
Land cover (2006 and 2016)	Land cover in 2006 and 2016 in the Sintang Regency	Ministry of Environment and Forestry (KLHK)
The road	National roads, provinces, districts, and sub-districts/villages	Geospatial Information Agency; scale 1: 50,000
The river	Big river	Geospatial Information Agency; scale 1: 50,000
Hub (central activity)	Hub location	Geospatial Information Agency; scale 1: 50,000
Mills	Location of mills	World Wildlife Fund (WWF)
Business use permits and location permits	Business use permits and location permits in the Sintang Regency	Business use permits from National Land Agency (BPN) and location permit from spatial planning (RTRW)
Forest cover 2016	Forest cover in 2016	Ministry of Environment and Forestry (KLHK)
Deforestation 2006–2016	Deforestation areas between 2006 and 2016	Ministry of Environment and Forestry (KLHK) analyzed by Hatfield Indonesia
Plantation 2006–2016	Plantation area between 2006 and 2016	Ministry of Environment and Forestry (KLHK)

6.3 Methodology

6.3.1 Land Projection Model

Land projection analysis was performed by comparing two scenarios: Business as Usual (BAU) and *Sintang Lestari* scenario (SL) or known as a green scenario. The definition of scenario planning was obtained from a series of discussions and workshops facilitated by the Conservation Strategy Fund (CSF) and the Sintang Regency Government.

The projection model is based on the analysis of historical land cover changes, transition potential, and factors driving land use change. Land use projections for both scenarios were projected until 2030, following the above scenario's planning targets.

1. *Scenario Development*
   The scenario was developed in collaboration with relevant stakeholders in Sintang through the Focus Group Discussion (FGD). Scenario planning was carried out through a multi-stakeholder process from October 1–5, 2018. The facilitators for this scenario planning were representatives from academic institutions, Non-Government Organizations, Civil Society Organizations, and government staff in Sintang with a total of 15 facilitators. Each institution sends representatives of five people. The facilitators were trained for 2 days by the CSF team on how the scenario planning process took place and how to facilitate these

Table 6.2 Key indicators for the SL model scenario

Indicator	Model description	Model application
Zero deforestation	Forest cover based on land cover in 2016 must be protected, and conversion is prohibited	Zero deforestation (applies to primary and secondary forests)
Sustainable utilization of natural resources without damaging the environment		
Forests are protected to maintain river water quality		
Protected forests are free from illegal farming and illegal logging		
Follow/respect rules of strategic environmental assessment (KLHS) and spatial planning (RTRW)	Cultivation areas are restricted. No planting is permitted outside the current concession area	Plantations are restricted

activities. The scenario planning was attended by 60 representatives from cross-institutional/stakeholder participants.

During the FGD process, the land projection model scenario was defined and agreed upon by the stakeholders. The definition of each modelling scenario is as follows:

(a) BAU scenario: This scenario is based on the current district planning, where each land allocation is defined as in the District Spatial Plan. Under this scenario, no restrictions are implemented for areas allocated as development areas. Forested land is categorized under development areas and is allowed to be cleared.

(b) SL scenario: This scenario is generated based on land cover in 2016 in which forest cover must be protected and conversion is forbidden, cultivation area is restricted and planting is not allowed outside the current concession area. Land cover predictions are generated with the following limitations:

- Zero deforestation: forest cover is considered as a protected area and no conversion of forest cover is allowed.
- Plantation restrictions: plantation is restricted outside the current concession area.

Limitations used in the SL scenario are taken from indicators in the policy direction matrix in the regional medium-term development plan (RPJMD) of the Sintang Regency for the 2016–2021 period. The planning documents have been aligned with sustainable district goals which are reflected through several indicators including forest and protected area conservation, air and environmental quality improvement, and others. Indicators included in the model are only those that can be displayed spatially. Furthermore, selected indicators were treated as a limiting factor for development areas in the modelling of the SL scenario as shown in Table 6.2.

2. *Projection of Land Use and Land Cover (LULC)*

We used the Land Change Modeler (LCM) module in TerrSet software to project the land cover. LCM predicts land change based on historical changes that occurred at two points in time. In this study, land cover in 2006 and 2016 is used as initial data to see trends in historical change. Land cover prediction is carried out for 2030. In general, there are three steps in LCM, namely, change analysis, transition potential, and change prediction, as follows (Eastman, 2021):

(a) The change analysis stage compares and analyzes land cover between two periods. At this stage, the most dominant land cover changes can be identified and mapped so that the driving factors for each change can be determined.
(b) The transition potential stage identifies and maps potential changes that may occur based on historical change data. Change projections can be made on one transition or multiple transitions based on the same driving factor. The change probability modelling is done using Multi-Layer Perceptron (MLP). MLP was chosen because it can model more than one change at the same time. Driving factor use was defined based on their influence on the transition. In this study, distance from the road, distance from the open area, and distance to existing plantation area are considered as the most influencing factors to land change in Sintang.
(c) Change projection is the last stage of the modelling process. At this stage, LCM projects land cover for the year 2030. The model determines how the variables influence future change and how much change took place between 2006 and 2016 and then calculate a relative amount of transition to the year 2030. To make the model more robust, Land Change Modeler allows for the incorporation of constraints or limiting factors; in this study forest cover on the year 2016 and plantation permits are treated as a limiting factor for deforestation and plantation expansion. The workflow for the model is shown in Fig. 6.2.

6.4 Results and Discussion

6.4.1 Land Use and Land Cover Changes (LULCC) in Sintang from 2006 to 2016

Land use and land cover (LULC) maps for the period 2006–2016 are used as a basis for mapping historical LULC changes in Sintang Regency. Between 2006 and 2016, the total LULCC in Sintang Regency was 226,311 ha, contributing to 10% of the total regency boundary (Fig. 6.3). Historical land cover change trends were obtained from land cover maps in 2006 and 2016. During this period, land cover change in Sintang Regency was dominated by plantation expansion (e.g., oil palm or rubber) (4.19%), secondary forest loss (−4.08%), and swamp shrub loss (0.98%). Further analysis showed that the conversion of the secondary forest was mostly to plantation (35%), shrubs (32%), and mixed dryland agriculture (29%). From the result, it can be seen that the increase in plantations occurred quite rapidly between the period of 2006 and 2016. This can be observed in the contribution of the Agriculture, Plantation, Forestry, and Fisheries sector, which reached 30% in 2011, followed

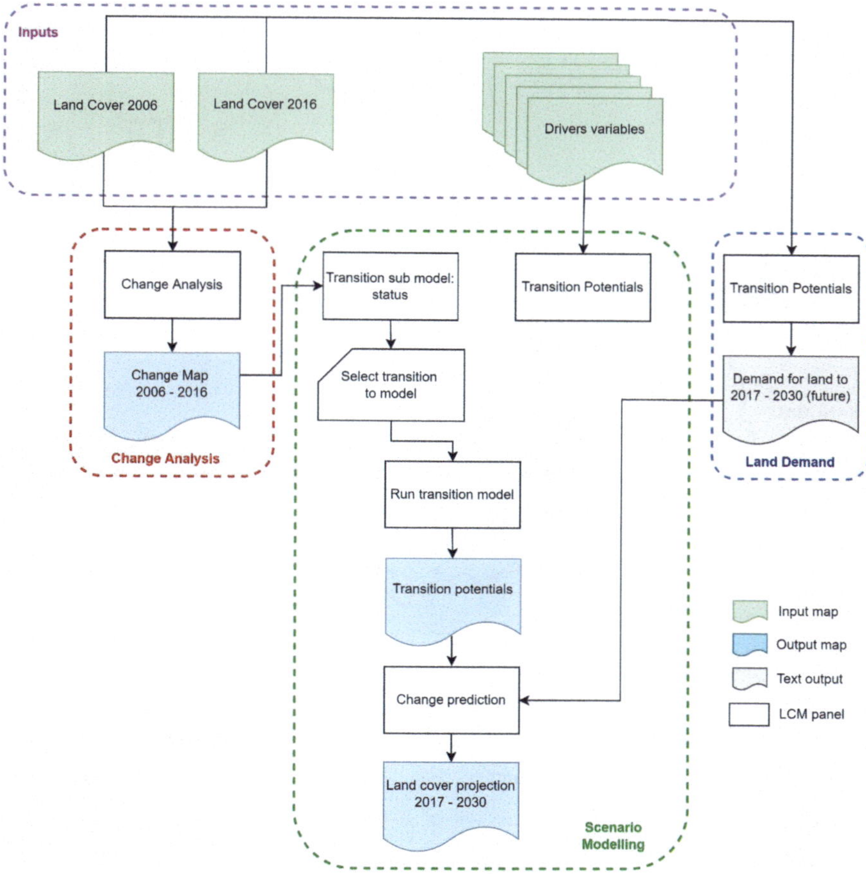

Fig. 6.2 The methodology used to analyze land use and land cover in Sintang using Land Change Modeler (LCM) software

by the Trade, Hotel, and Restaurant sector (20%) and other sectors (Central Bureau of Statistics, 2012).

This is also supported by the increase in the number of plantation crops, especially in oil palm plantations. The increase in the number of plantations in Sintang is almost the same as the decrease in the number of deforested areas in secondary forests. This means that the process of changing land status from forest to plantation has begun by changing from secondary forest to plantation. The main contributor to the increase in plantations during this period in Sintang Regency was secondary forest conversion. This trend is expected to be continued in the future, especially for oil palm plantations to meet the production target set by the government.

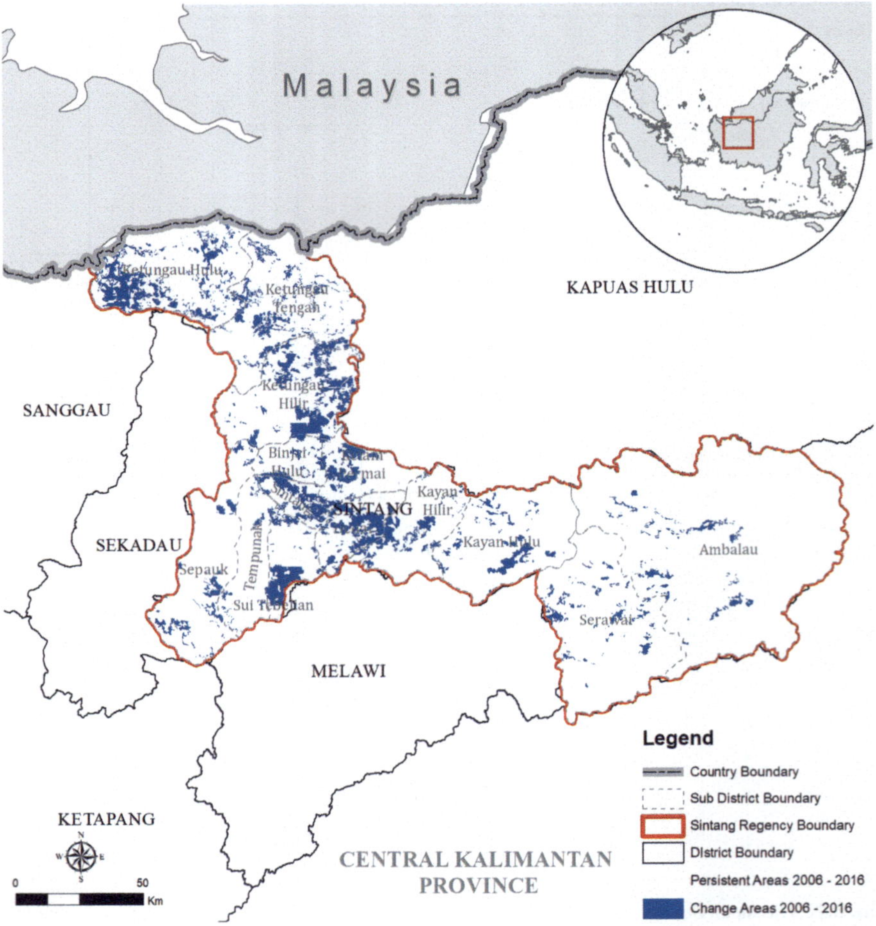

Fig. 6.3 Land use and land cover changes in Sintang from 2006 to 2016. (Source: Map produced based on the analysis)

6.4.1.1 Projection of Changes in Land Use and Land Cover from 2016 to 2030

BAU Scenario

The result of the BAU scenario in Fig. 6.4 is generated using Land Change Modeler Software. Under the BAU scenario, the software projected the future land cover based on the information from historical changes that occurred within the area. There are no limiting factors applied in the scenario. The result showed that, under the BAU scenario, by the year 2030 the regency expects to continue secondary forest loss (107,000 ha) and primary forest loss (10,000 ha). Plantation and open land are expected to be increased continuously to meet the government's development target

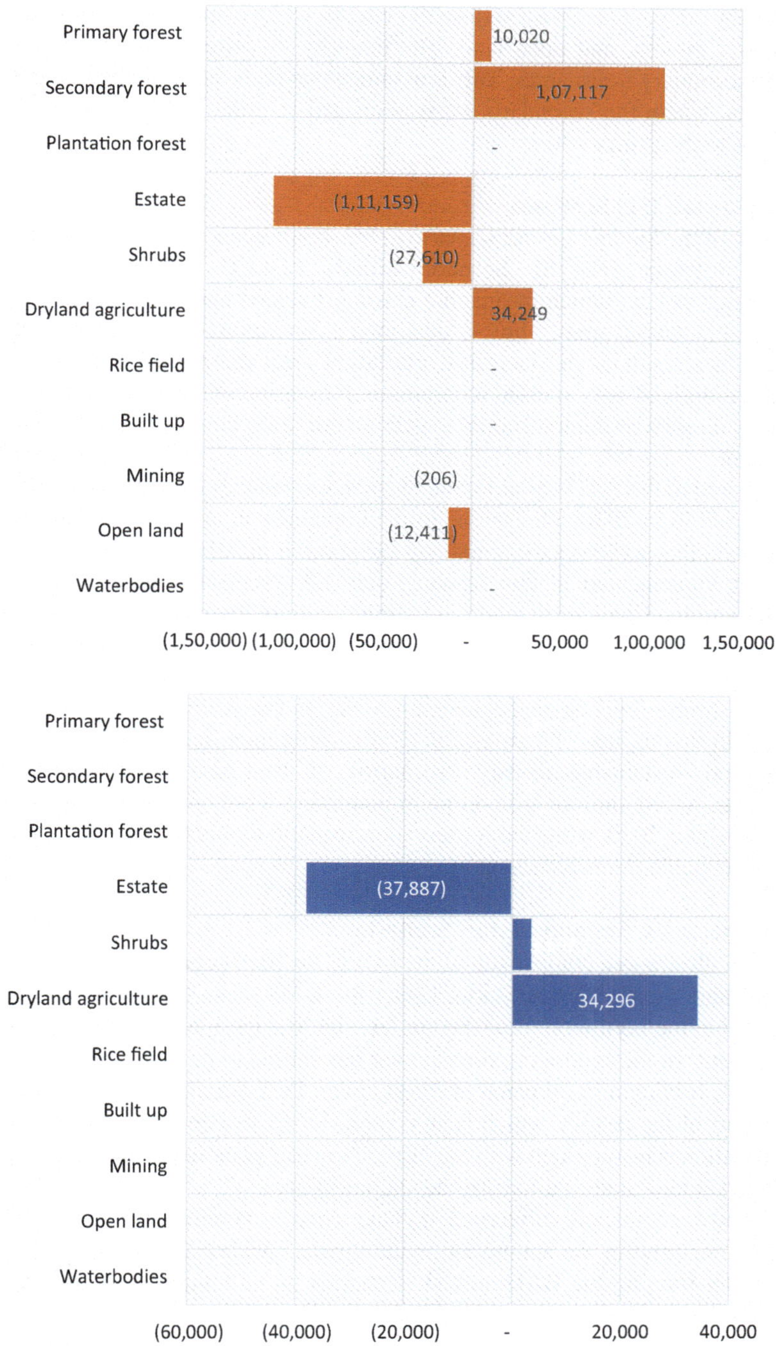

Fig. 6.4 Land cover projection under BAU and *Sintang Lestari* scenario from 2016 to 2030

of 2030. By 2030, some forest areas will be replaced by plantations, shrubs, dryland agriculture, mining, and open land. The majority of deforestation is expected to occur in secondary forest areas. This is assumed due to the accessibility factor of the forest area, where the secondary forest is located in a flat and accessible location compared to the primary forest area.

Sintang Lestari (SL) Scenario

The result of the Sintang Lestari scenario in Fig. 6.4 is generated using Land Change Modeler Software. The SL Scenario expects LULCC to implement a zero-deforestation policy, which means that existing forested land will remain the same in 2030. Plantation areas, shrubs, and mixed agricultural land are expected to change. As a result of the increased plantation area, the agricultural area, scrub, and mixed dryland are expected to decrease. Expansion of plantation areas will be limited as a result of implementing constraints for areas outside the current cultivation permit.

Compared to the BAU scenario, the SL scenario reduces plantation expansion by more than 50% in 2030. This is due to the implementation of a zero-deforestation commitment that preserves the remaining forest cover in 2016. Moreover, the second constraint implemented in this scenario only allows plantation expansion under existing permits. Changes in land cover from 2006 to 2016 showed that secondary forest is the biggest contributor to the increase in plantations, so this limits the massive plantation expansion. However, plantation land is still experiencing expansion from shrubs (3592 ha) and conversion to dryland agriculture (34,295 ha). This is in line with finding from Sharma et al. (2019) using three scenarios (BAU, conservation, and sustainable-intensive scenario) of land use projection in West Kalimantan which showed that oil palm plantation will continue increasing up to 3.2 million ha in 2030, while the conservation scenario is only 0.53 and 2.8 million ha under sustainable-intensive scenario.

BAU Scenario Vs. Sintang Lestari Scenario

Figure 6.5 illustrates a comparison of the LULC for 2016 conditions, the BAU 2030 scenario, and the SL 2030 scenario. Under the BAU scenario, forest cover almost disappears in the northwest, near the border with Malaysia, as a result of plantation development. In the southwest, forest cover has shifted to mixed scrub and dryland agriculture, leaving a small patch of forest cover. Deforestation is also found in the eastern part of the district, which is an ecotone area between forest and non-forest. This condition is inconsistent with the district's spatial plan, where the eastern part of the district is allocated as a forestry development area.

The most significant difference in land change between the BAU and SL scenarios occurred in the southwest of the district. Additional information can be obtained by running the BAU and SL scenarios on an annual basis to determine annual trends in LULC changes. In both scenarios, the results showed a similar pattern for changes in the plantation area. Under the BAU scenario, plantations are expected to grow and then decline gradually. However, the SL scenario revealed no changes in primary forest, secondary forest, open land, or mining.

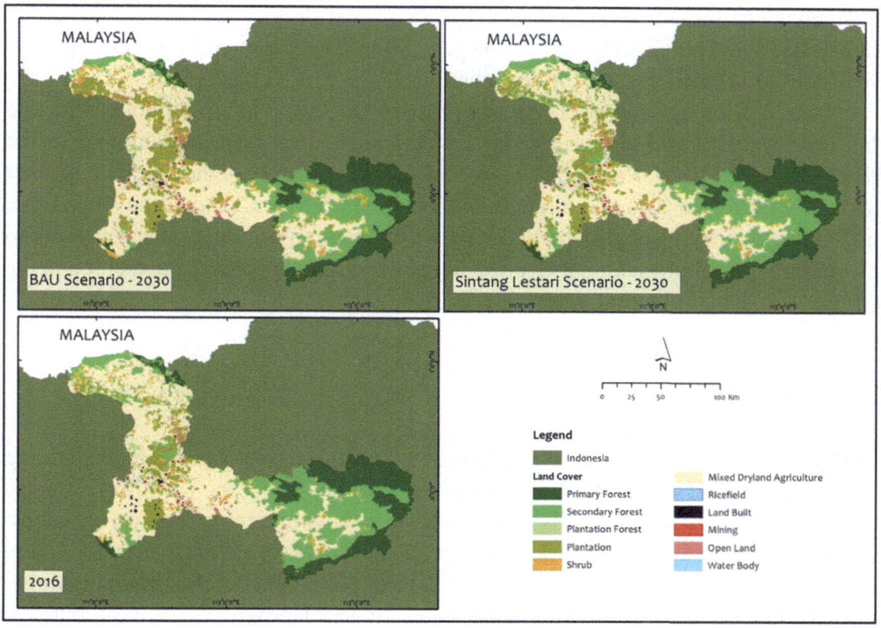

Fig. 6.5 Comparison of the LULC for 2016, the BAU 2030, and the SL 2030 scenarios

Generally, the total change in the area of Sintang Regency from 2006 to 2016 is 226,311 ha or 10% of the regency area. The high rate of deforestation in secondary forests is due to Sintang's economic structure, which still relies on natural resources and land. Deforestation also occurs in bushland and primary forest, but at a slower rate than in the secondary forest. Previous studies discovered that the status of the deforested land area had changed its function into plantations, shrubs, and dryland agriculture (Sukri et al., 2020). Meanwhile, the number of plantation crops, particularly oil palm plantations, has increased. The increase in the number of plantations in Sintang is almost the same as the deforested areas in secondary forests. This means that the process of converting land from forest to plantation has started by converting secondary forest to the plantation. Our findings were consistent with previous research indicating the rapid expansion of plantations, particularly oil palm, in Kalimantan (Carlson et al., 2012; Siregar et al., 2018). From 1992 to 2012, the pattern of forest conversion in West Kalimantan Province started with forest conversion to other land use (mostly rain-fed paddy fields or shrubs), followed by conversion to the plantation as the final land use (Siregar et al., 2018).

6.4.1.2 Further Analysis and Comparison with More Recent Landscape Planning and Sintang Lestari Regulations Is Needed

The scenarios presented in this analysis are not based on the most recent data, nor are they compared to the most recent Sintang landscape planning and policy documents. However, by the time the scenario was completed, the government of Sintang was made aware of these results, and open discussions with relevant agencies (viz., the Landscape and Spatial Planning Agency, and the Development Planning Agency) were held to inform on their further planning.

By 2022, Sintang plans to revise its primary landscape planning document (the Regional Spatial Plan) and also to have a detailed sectoral spatial planning (or the *Rencana Detail Tata Ruang*). The scenarios presented here need to be further updated with recent data and aligned with Sintang priorities. And at the same time, the leading agency of landscape planning would need to use the updated version to inform their decision-making as well as to further align the trade-off presented in the scenarios to actual planning.

6.5 Conclusion

Land use in Indonesia is very dynamic and changes over time. Understanding landscape dynamics and ongoing processes within the landscape requires an understanding of land use change. It is used to assist the government in making decisions. Understanding historical land use change can also serve as a strong basis for future land use projects. This is to comprehend the consequences of current planning practices and to design government interventions in a landscape for future benefits.

Based on the analysis of land cover change from 2006 to 2016, the secondary forest area experienced the largest decrease (−87,680 ha), while the plantation area had the largest increase (87,540 ha). Further analysis showed that most of the secondary forest was converted to plantations. This trend is projected to continue under the BAU scenario in which plantations will grow to 253,571 ha in 2030. The SL scenario (green scenario) predicted only 180,300 ha of plantation growth by 2030. However, the projection results for the SL scenario used "zero deforestation," indicating that significant plantation development can still be achieved while protecting the remaining forest cover in Sintang Regency.

If the government chooses the SL scenario, it can still maintain economic growth, but it will require significant efforts to improve the value added of current forest and land-based resources, increase the current productivity of agricultural and plantation products, and shift the GDP main sectors from land-based resources to other sectors such as service and financial sectors, communication, and others to generate GDP.

However, limiting the expansion of plantation areas is only possible if current spatial plans are monitored and enforcement on non-compliance with spatial planning is implemented. The trade-offs should also be acknowledged and described clearly in the forthcoming planning documents.

Acknowledgements This research is part of the author project on Sustainable Landscape Planning in Sintang, West Kalimantan, Indonesia. The authors would like to express their sincere gratitude to the David and Lucile Packard Foundation for funding this project. The authors would also like to acknowledge the support from Hatfield Indonesia, an institution that provides training and analysis during the implementation of this project.

References

Abram, K. N., Meijaard, E., Wilson, A. K., et al. (2017). Oil-palm community conflict mapping in Indonesia: A case for better community liaison in planning for development initiatives. *Applied Geography, 78*, 33–44. https://doi.org/10.1016/j.apgeog.2016.10.005

Afriyanti, D., Hein, L., Kroeze, C., et al. (2019). Scenarios for withdrawal of oil palm plantations from peatlands in Jambi Province, Sumatra, Indonesia. *Regional Environmental Change, 19*, 1201–1215. https://doi.org/10.1007/s10113-018-1452-1

Aznar-Sánchez, J. A., Belmonte-Ureña, L. J., López-Serrano, M. J., et al. (2018). Forest ecosystem services: An analysis of worldwide research. *Forests, 9*(8), 453. https://doi.org/10.3390/f9080453

Brockerhoff, E. G., Barbaro, L., Castagneyrol, B., et al. (2017). Forest biodiversity, ecosystem functioning and the provision of ecosystem services. *Biodiversity and Conservation, 26*(13), 3005–3035. https://doi.org/10.1007/s10531-017-1453-2

Carlson, K. M., Curran, L. M., Ratnasari, D., et al. (2012). Committed carbon emissions, deforestation, and community land conversion from oil palm plantation expansion in West Kalimantan, Indonesia. *Proceedings of the National Academy of Sciences, 109*, 7559–7564.

Central Bureau of Statistic Indonesia. (2020). *Kabupaten Sintang Dalam Angka 2019*. Badan Pusat Statistik. Retrieved from www.sintang.bps.go.id

Central Bureau of Statistics. (2012). Retrieved from https://www.bps.go.id

Eastman, J. R. (2021). *TerrSet 2020 geospatial monitoring and modelling system. Manuals*. Clarklab.

Fankhauser, S., & Tol, R. S. J. (2005). On climate change and economic growth. *Resource and Energy Economics, 27*, 1–17. https://doi.org/10.1016/j.reseneeco.2004.03.003

García-Nieto, A. P., García-Llorente, M., Iniesta-Arandia, I., et al. (2013). Mapping forest ecosystem services: From providing units to beneficiaries. *Ecosystem Services, 4*, 126–138. https://doi.org/10.1016/j.ecoser.2013.03.003

Gunarso, P., Hartoyo, M. E., Agus, F., & Killeen, T. J. (2013). *Oil palm and land-use change in Indonesia, Malaysia and Papua New Guinea*. Reports from the Technical Panels of the 2nd Greenhouse Gas Working Group of the Roundtable on Sustainable Palm Oil (RSPO). RSPO.

Hwang, J. H., & Yoo, S. H. (2012). Energy consumption, CO_2 emissions, and economic growth: Evidence from Indonesia. *Quality & Quantity, 48*(1), 63–73. https://doi.org/10.1007/s11135-012-9749-5

Indonesia Geospatial Agency. (2006). Retrieved from https://tanahair.indonesia.go.id/portal-web/

Knight, K. W., & Schor, J. B. (2014). Economic growth and climate change: A cross-national analysis of territorial and consumption-based carbon emissions in high-income countries. *Sustainability, 6*, 3722–3731. https://doi.org/10.3390/su6063722

Laurance, W. F., Sayer, J., & Cassman, K. G. (2014). Agricultural expansion and its impacts on tropical nature. *Trends in Ecology & Evolution, 29*(2), 107–116. https://doi.org/10.1016/j.tree.2013.12.001

Malahayati, M., & Masui, T. (2019). The impact of green house gas mitigation policy for land use and the forestry sector in Indonesia: Applying the computable general equilibrium model. *Forest Policy and Economics, 109*, 102003. https://doi.org/10.1016/j.forpol.2019.102003

Margono, B. A., Potapov, P. V., Turubanova, S., et al. (2014). Primary forest cover loss in Indonesia over 2000–2012. *Nature Climate Change, 4*, 1–6. https://doi.org/10.1038/NCLIMATE2277

Noorbakhsh, F., & Ranjan, S. (1999). A model for sustainable development: Integrating environmental impact assessment and project planning. *Impact Assessment and Project Appraisal, 17*(4), 283–293. https://doi.org/10.3152/147154699781767684

Nowak, D. J., Poudyal, N. C., McNulty, S. G. (2015). Chapter 4: Forest ecosystem services: Carbon and air quality. In *Trees at work: Economic accounting for forest ecosystem services in the U.S. South*. Retrieved from https://www.srs.fs.usda.gov/pubs/gtr/gtr_srs226/gtr_srs226_ch4.pdf

Pearce, D. W., Atkinson, G. D., & Dubourg, W. R. (1994). The economics of sustainable development. *Annual Review of Energy and the Environment, 19*, 457–474.

Schebek, L., Mizgajski, J. T., Schaldach, R., & Wimmer, F. (2018). Land-use change and CO_2 emissions associated with oil palm expansion in Indonesia by 2020. In B. Otjacques, P. Hitzelberger, S. Naumann, & V. Wohlgemuth (Eds.), *From science to society. Progress in IS*. Springer.

Shahbaz, M., Hye, Q. M. A., Tiwari, A. K., & Leitão, N. C. (2013). Economic growth, energy consumption, financial development, international trade and CO_2 emissions in Indonesia. *Renewable and Sustainable Energy Reviews, 25*, 109–121.

Sharma, S. K., Baral, H., Laumonier, Y., Okarda, B., Komarudin, H., Purnomo, H., & Pacheho, P. (2019). Ecosystem services under oil palm expansion scenarios in West Kalimantan, Indonesia. *Ecosystem Services, 39*. https://doi.org/10.1016/j.ecoser.2019.100978

Siregar, P. G., Supriatna, J., Koestoer, R. H., & Harmantyo, D. (2018). System dynamics modeling of land use change in West Kalimantan, Indonesia. *BIOTROPIA-The Southeast Asian Journal of Tropical Biology, 25*(2), 103–111. https://doi.org/10.11598/btb.2018.25.2.792

Sukri, W., Oktavianus, B., Anastasia, F. P., Prawiyanto, Y., Isnandar, A., Subariyanti, E., Hermanto, F., Epi, P., Perdana, Z. H., Susanti, S., Setiawan, H., Putri, S. C., Wihastuti, S. J., Pasaribu, P. R., Istanti, R. P., & Peteru, S. (2020). Sintang District, West Kalimantan, Indonesia. In C. Stickler et al. (Eds.), *The state of jurisdictional sustainability*. EII/CIFOR.

Land Use and Land Cover Change Prediction Using ANN-CA Model

7

K. Norizah, J. Jamhuri, M. Balqis, I. Mohd Hasmadi, and A. A. Nor Akmar

Abstract

Land cover changes are characterised by the loss of natural resources, the loss of forests to urban development or the loss of agricultural areas to urban growth. This study applies the artificial neural network (ANN) method in cellular automata (CA) modelling to predict the changes, pattern and transition matrix of land use and land cover (LULC) changes for three decades with 10-year intervals: 1995–2005, 2005–2015 and 2015–2025. The maps for 1995, 2005 and 2015 obtained from Landsat 8 images are classified using the Maximum Likelihood Classification (MLC) algorithm with overall accuracies of 86%, 84% and 80%, respectively. Then, the prediction of LULC for 2025 is made after using the real map to satisfactorily validate the predicted map for 2015 derived from the map of 1995–2005, with the correctness percentage of 82.86 and the overall Kappa coefficients of 0.75. The forest cover shows a high decrement in the first 10-year period (44.87%). This trend is in line with the increasing areas of LULC for oil palm cultivation and urban development until 2025. Stable transitions can

K. Norizah (✉)
Institute of Tropical Forestry and Forest Products (INTROP), Universiti Putra Malaysia (UPM), Serdang, Malaysia

Department of Forestry Science and Biodiversity, Faculty of Forestry and Environment, Universiti Putra Malaysia (UPM), Serdang, Malaysia
e-mail: norizah_k@upm.edu.my

J. Jamhuri · M. Balqis · I. Mohd Hasmadi
Department of Forestry Science and Biodiversity, Faculty of Forestry and Environment, Universiti Putra Malaysia (UPM), Serdang, Malaysia

A. A. Nor Akmar
Department of Nature Parks and Recreation, Faculty of Forestry and Environment, Universiti Putra Malaysia (UPM), Serdang, Malaysia

Z. Samdin et al. (eds.), *Tropical Forest Ecosystem Services in Improving Livelihoods For Local Communities*, https://doi.org/10.1007/978-981-19-3342-4_7

be observed from the ANN-CA model transition matrix, with values of 0.644, 0.541 and 0.787 transitions in 1995–2005, 2005–2015 and 2015–2025, respectively, for the forest area. For the urban area, the transition has been observed as 0.595, 0.736 and 0.954 in 1995–2005, 2005–2015 and 2015–2025, respectively. Meanwhile, oil palm only shows a stable transition of land-use changes in the second 20-year period, 2005–2015 with 0.606 and 2015–2025 with 0.843. The simulation of LULC changes in this study using the ANN-CA modelling can be a useful decision-making support tool to gauge the sustainability of urban area expansion, agricultural cultivation and production, as well as conservation of the remaining forest areas.

Keywords

Land cover · Artificial neural network · Cellular automata · Land use and land cover · Agricultural cultivation and production

7.1 Introduction

Land use and land cover changes (LULC) in Malaysia are primarily attributed to the expansion of oil palm plantation and urbanisation (How Jin Aik et al., 2020; Wan Mohd Jaafar et al., 2020; Zulfa et al., 2020; Hua, 2017; Aburas et al., 2015). These changes will have grave consequences for the natural ecosystems, for example, loss of natural areas, including forests (How Jin Aik et al., 2020; Wan Mohd Jaafar et al., 2020; Zulfa et al., 2020), driving global climate change (How Jin Aik et al., 2020; Wan Mohd Jaafar et al., 2020) as well as degradation of ecosystems and biodiversity (Mallick et al., 2021; Wan Mohd Jaafar et al., 2020). However, the changes of LULC do not always produce negative impacts. There are benefits such as biodiversity richness (Foster et al., 2011; Azhar et al., 2015; Teuscher et al., 2016), including floras and faunas in forest-converted oil palm plantations (Adila et al., 2017; Sasidhran et al., 2016; Teuscher et al., 2016; Foster et al., 2011), and well-planned and purpose-designed urban landscapes (Tee et al., 2018; Aida et al., 2016).

Beyer et al. (2020) and Cumming et al. (2014) suggested that the exploitations of natural ecosystems are due to the interaction and mutually reinforcing processes of technological advances, population growth and urbanisation; and these factors have impacts on the sustainability of resource usage. Here is an example of the contribution of this phenomenon. Although urban populations still rely on the natural ecosystem, the priority of socio-economic activities causes an increase in the urban population size and urban expansion (Marzuki & Jais, 2020; Egidi et al., 2020; Salvati & Lamonica, 2020; Hua & Ping, 2018). As the population grows, there are changes in the socio-economic variables and demand for necessities of life, such as food, fibre and fuel. This has been observed from the progress and development of economic activities in Malaysia. According to Parveez et al. (2021), palm oil is vital for the national economy and an important commodity in the international edible oil market as Malaysia is the world's second-largest producer. The growth of this plantation sector causes acreage expansion for oil palm cultivation and the application of technology for sustainable productions and yields (Abdul-Hamid et al., 2020).

Malaysia has been developing rapidly over the past 30 years as various industrial and business sectors embarked on expansions to generate higher volumes of goods and services (Miyamoto et al., 2014). The demand for land increases as the population grows and more advanced technological establishments are set up. Statistics obtained from the Global Forest Watch (2020) website https://www.globalforestwatch.org/dashboards/country/MYS/ show vegetation was Malaysia's prominent land cover in 2015: forested areas covered 20.2 million ha; agriculture including oil palm plantation, 10.8 million ha; wetland, 1.35 million ha; and settlements, 382 kha. As Malaysia is a developing country, forested lands are often being cleared for various development purposes (Kanniah & Ho, 2017; Kanniah et al., 2015; Hasan & Nair, 2014). In most parts of Malaysia, the size of non-vegetation land cover has been increasing, and the forested land area was decreasing, due to urban growth and oil palm acreage expansion (Asnawi et al., 2018; Sheikhi & Kanniah, 2018; Hua & Ping, 2018; Abdullah & Nakagoshi, 2008). With strong economic growth and fast-paced developments, regions with highly populated areas are experiencing forest fragmentation (Reza, 2014).

Change detection analysis using remotely sensed imagery data has been conducted worldwide and reported with varying degrees of accuracy. The existing analysis and algorithm model for change detection analysis as well as various types of available imagery might be the reason for this trend. For example, Mahamud et al. (2019) employed the Cellular Automata-Markovian (CA-Markov) model to predict land use and land cover changes in the year 2025 based on two land-use maps of 1991 and 2000; they then calculated the transition matrix of the cell that was expected to change the land use class in the future. Hua (2017) used the CA-Markov model for simulating the water quality of the Malacca River, and it provides the planners with a system of early warning by predicting the trend and consequences of changes in the watershed. A similar concept of LULC prediction was also studied by Mallick et al. (2021), Abbas et al. (2021) and Saputra and Lee (2019), who used the artificial neural network-cellular automata model (ANN-CA). All the research mentioned above provide future management and mitigation measures to the planners, managers and stakeholders while safeguarding the sustainability of the environment.

For LULC study and analysis, mapping and regular monitoring are crucial as they play a significant role in planning, management and sustainable development at various levels. Hence, this chapter examines the dynamics of LULC changes using the CA-Markov model analysis. The choice of this model is based on the following requirements: dynamic simulation capability, high efficiency with data scarcity, simple calibration, ability to simulate multiple land covers and complex patterns, which sheds light on the changes over time, transition and the current landscape. By understanding the dynamics of LULC, stakeholders, managers and other responsible agencies could analyse the previous management's decisions and gain insight into the potential consequences of the current decisions before implementing them for sustainable management in the future.

7.2 Materials and Methods

7.2.1 Study Area

The prediction of LULC was conducted for the Selangor state; centroid coordinate, 3.0738°N, 101.5183°E, including the largest city of Malaysia, Kuala Lumpur, and centroid coordinate, 3.1390°N, 101.6869°E (Fig. 7.1). This state is characterised by a significant amount of annual rainfall, with 2960 mm precipitation and an average temperature of 26.1 °C. Kuala Lumpur City, located in some parts of Selangor as well as Petaling, experiences an extreme extent of man-made activities due to a large number of developments and higher population density (Asnawi et al., 2018). Along the coastline in the west and the northern region of Selangor, the lands are characterised mainly by oil palm plantations and paddy fields. In addition, Selangor is endowed with a unique forest ecosystem, with mangroves in the western region, peat swamp in an area stretching from the west to north of Selangor and hill forest in the eastern inland region.

7.2.2 Methodology

The conceptual framework for detecting and predicting the changes of LULC is outlined in Fig. 7.1. The process includes the following key steps: (1) data acquisition and preparation; (2) image classification using the maximum likelihood

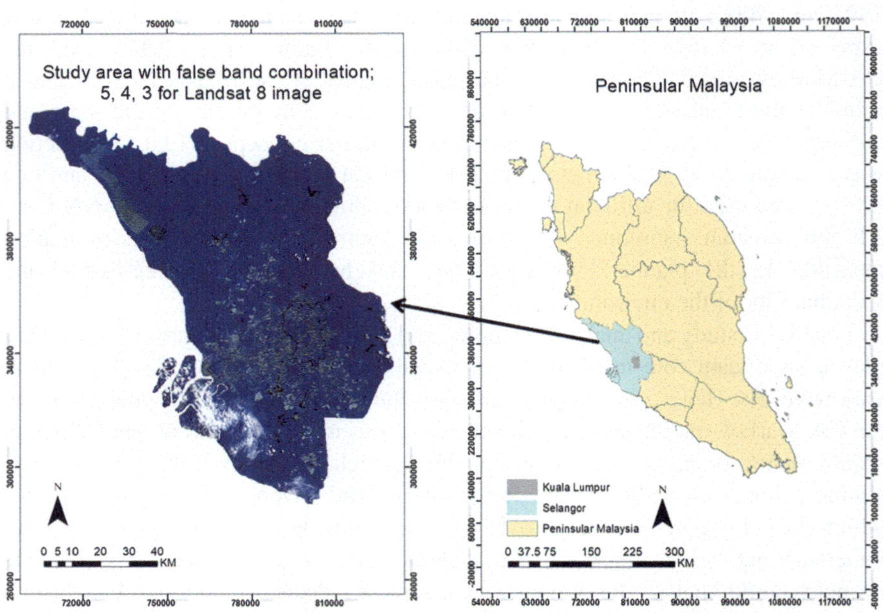

Fig. 7.1 Location of the study area showing Selangor and Kuala Lumpur City

classification (MLC) for images of 1995, 2005 and 2015; (3) LULC changes analysis; and (4) simulation of LULC in 2025 and 2035 using the ANN-CA model.

7.2.2.1 Data Acquisition and Preparation

To identify the time-series forest cover changes, three types of remotely sensed imagery data are used, namely, Landsat 4–5, Landsat 7 and Landsat 8. Two scenes were required to present the study area; path/row: 127/57 and 127/58. Details of the satellite data used in this study are given in Table 7.1. These data were downloaded for free from the US Geological Survey (USGS) website https://earthexplorer.usgs.gov/.

Since the imageries obtained from the USGS website already have the spatial reference, they are automatically used as the coordinate system: Kertau Rectified Skew Orthomorphic (RSO) Malaya Meter. The scenes of all the imageries are then mosaicked. Subsequently, a subset procedure is performed to retrieve the study area. All the image pre-processing is conducted in QGIS Desktop software version 3.18.1 (https://qgis.org/en/site/).

7.2.2.2 Image Classification and Accuracy Assessment

Li et al. (2014) discuss several image classification techniques, including pixel-based, sub-pixel-based and object-based. They describe pixel-based as a common pixel-wise classification technique for LULC detection. To detect the LULC in this study, we use supervised classification with the MLC algorithm. The training samples are created based on six LULC classes, which are (1) forest, (2) oil palm, (3) urban area, (4) paddy field, (5) water bodies and (6) open area (Table 7.2).

Accuracy assessment is conducted in the later stage to examine the performance of classification. As the procedure for ground-truthing could be more costly and time-consuming, the accuracy assessment is done by comparing the classified image with another data source that is considered as accurate with the ground-based truth data, the Google Earth (GE) imagery (https://earth.google.com/web/) (Atiqah et al., 2020; Thomas et al., 2017).

Accuracy of the supervised classification derived map is conducted by examining the LULC at each generated validation point and compared with the GE imagery time series (Atiqah et al., 2020; Thomas et al., 2017) using the same points.

To indicate the agreement of the classification conducted with the comparison to the GE, Cohen's (1960) kappa was used, and the coefficient can be interpreted as,

Table 7.1 The characteristics of the satellite imageries

Year	Satellite image	Acquisition date	Resolution (m)	Band combination
1995	Landsat 4–5	15 Jan 1995	30	4 (NIR: 0.77–0.90 μm) 3 (RED: 0.63–0.69 μm) 2 (green: 0.52–0.60 μm)
2005	Landsat 7	4 Dec 2005	30	4 (NIR: 0.77–0.90 μm) 3 (RED: 0.63–0.69 μm) 2 (green: 0.52–0.60 μm)
2015	Landsat 8	3 Sept 2015	30	5 (NIR: 0.85–0.88 μm) 4 (RED: 0.64–0.67 μm) 3 (green: 0.53–0.59 μm)

Table 7.2 Descriptions of land use and land cover class

Land use and land cover class	Description
1. Forest	Areas covered with native trees, where no human activity can be detected
2. Oil palm	Areas covered with oil palm activity with star-shaped features from the image
3. Urban area	Areas with developed buildings or environments such as residential area, city area or industrial area
4. Paddy field	Flooded fields are used to cultivate rice; the area is either found dry or wet depending on the season
5. Waterbodies	Areas covered with rivers, lakes or dams
6. Open area	Vacant areas are covered with soil

viz. (1) ≤ 0 indicating no agreement, (2) 0.01–0.20 as none to slight, (3) 0.21–0.40 as fair, (4) 0.41–0.60 as moderate, (5) 0.61–0.80 as substantial and (6) 0.81–1.00 as almost perfect agreement. Accuracy for supervised classification derived map was conducted by examining the LULC at each generated validation point and compared to time series of GE imagery using the same points. A total of 100 validation points were distributed over the study area randomly with a p value of 0.0538 and Global Moran's Index of 0.0493 (Appendix 1).

7.2.2.3 ANN-CA Model Analysis and Validation

The application of the ANN-CA model analysis is conducted in the QGIS software with the MOLUSCE plugin. The spatial variable input required for this analysis is the digital elevation model (DEM) of the study area and the distance to the road (Appendix 2). The DEM was extracted from an Advance Land Observing Satellite-Phased Array type L-band Synthetic Aperture Radar (ALOS PALSAR) image with 12.5 m resolution.

For validation, the real LULC map is compared with the predicted map for 2015 (Appendix 3). The kappa coefficient is used to measure the actual agreement between both the real and the predicted map (Abbas et al., 2021; Saputra & Lee, 2019). The simulation results for the spatial variables show a correctness percentage of 82.86 and an overall kappa coefficient of 0.75 (Appendix 4). Based on Cohen's (1960) kappa, these two spatial variables, namely, DEM and distance to roads, can be interpreted as having substantial impacts on the transition matrix of the six LULC categories.

7.3 Results

7.3.1 Land Use and Land Cover Analysis

Figure 7.2a–c illustrates the LULC maps derived from the MLC classifier algorithm of the study area in 1995, 2005 and 2015, respectively. The accuracy agreement of the LULC maps of 1995, 2005 and 2015 as measured with the kappa coefficient is

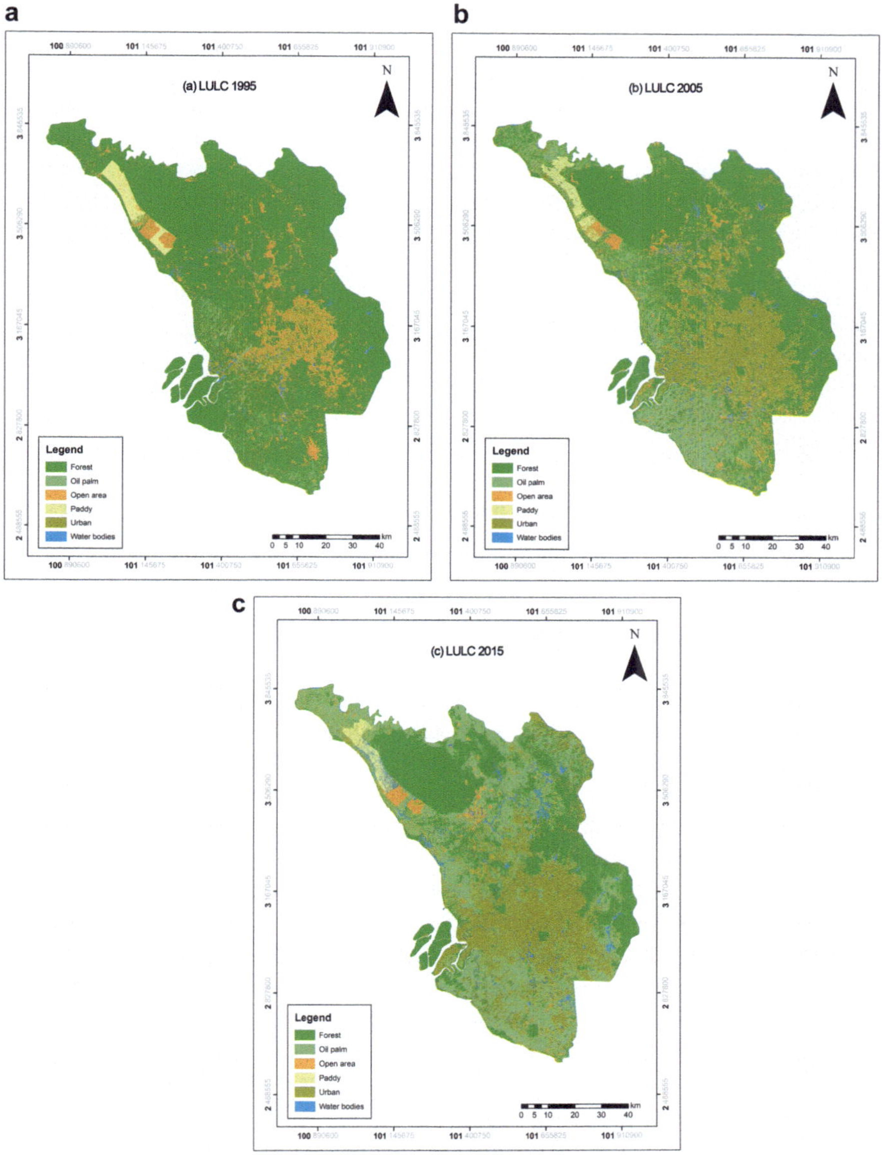

Fig. 7.2 LULC classification with MLC classifier algorithm for (**a**) 1995, (**b**) 2005 and (**c**) 2015

presented in Table 7.3. The LULC map for 1995 has a kappa coefficient value of over 0.81; this indicates, in comparison with GE, the accuracy of the LULC produced from the MLC classifier algorithm is almost perfect (Cohen, 1960) with an overall classification accuracy of 86%. Meanwhile, for the LULC maps of 2005 and 2015, the accuracies are fairly good compared to the GE since the kappa

Table 7.3 Accuracy assessment maps of 1995, 2005 and 2015

LULC classes	1995		2005		2015	
	Producer's accuracy	User's accuracy	Producer's accuracy	User's accuracy	Producer's accuracy	User's accuracy
Forest	87.50%	94.92%	92.16%	87.04%	89.80%	89.80%
Oil palm	70.00%	58.33%	72.72%	50.00%	76.47%	68.42%
Open area	100.00%	85.71%	70.00%	87.50%	100.00%	87.5%
Paddy	50.00%	50.00%	60.00%	75.00%	60.00%	100.00%
Urban	100.00%	72.73%	83.33%	81.25%	53.33%	53.33%
Water bodies	87.5%	100%	80.00%	100%	71.43%	83.33%
Overall accuracy	86%		84%		80%	
Kappa coefficient	0.81		0.76		0.73	

coefficient values are less than 0.80 and the overall accuracies are 84% and 80% for images 2005 and 2015, respectively.

7.3.2 Land-Use Land-Cover Transition Analysis

The LULC maps of 1995 and 2005 are used to predict the LULC of 2015. The satisfactory results from the model validation of the predicted LULC for 2015 are later used to predict the LULC of 2025 with LULC of 2005. The predicted LULC map of 2025 is presented in Fig. 7.3, and the predicted changes of LULC of 2025 are presented in Table 7.4. Abbas et al. (2021) explained that the temporal changes within a set of LULC categories could be observed through the transition matrix. Tables 7.5, 7.6 and 7.7 present the temporal changes for the 10-year periods between 1995 and 2005, 2005 and 2015 and 2015 and 2025. The matrix rows represent the LULC in the initial year, and the columns represent the LULC in the final year with the same order. The transition from the initial year to the final year can be observed through the crosswise values. The diagonal value entries show the size of class stability; values close to 1 represent the stability of the LULC category.

The forest shows continual decrement in land cover from 1995 to 2025. A high decrement is observed in the first 10-year period, 1995–2005 with 44.87%; then the changes slow down in the next two periods to 20.95% and 5.45% in 2005–2015 and 2015–2025, respectively. The changes of forest cover show consistent stabilities with values of 0.644, 0.541 and 0.787 transitions in 1995–2005, 2005–2015 and 2015–2025, respectively. While the forest cover recorded gradual decrement over the 30 years, the lands used for oil palm cultivation and urban development show increments. The transition value for oil palm is stable in the second and the third 10-year period, 0.606 and 0.843, respectively. The transitions of the urban area are almost stable in the 30 years with values of 0.595, 0.736 and 0.954 in 1995–2005,

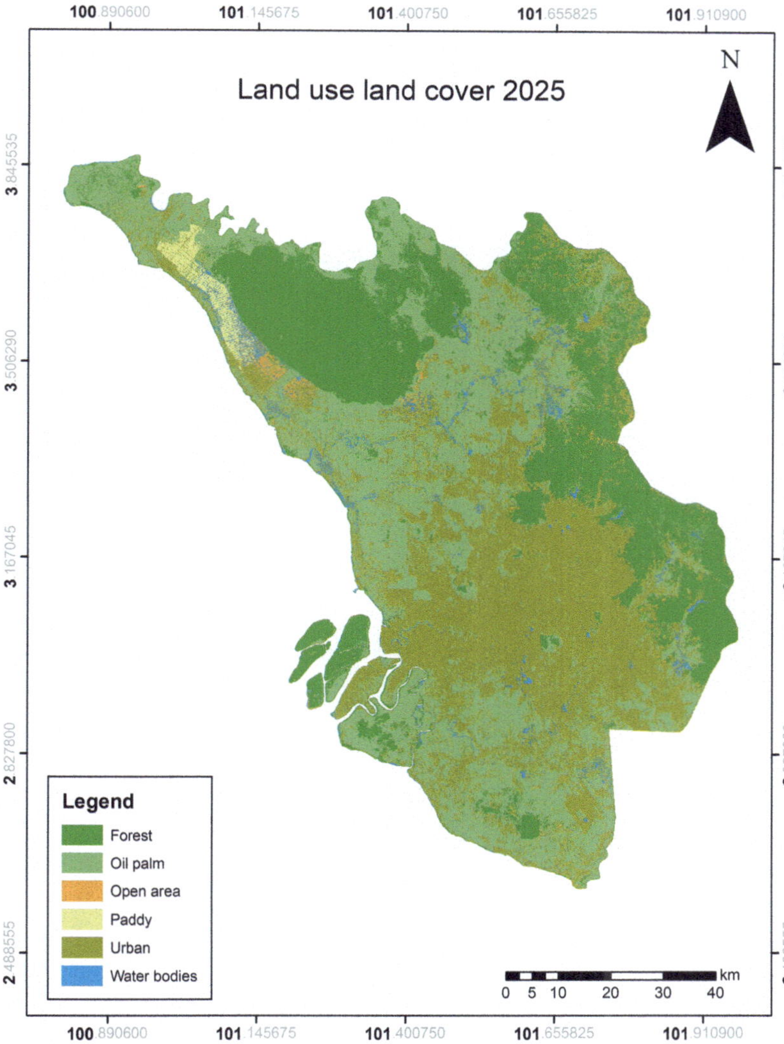

Fig. 7.3 LULC simulation and prediction using ANN-CA model for 2025 derived from LULC map of 2005 and 2015

2005–2015 and 2015–2025, respectively. The land use of paddy and the land cover of water bodies initially show an increment in the first 20-year period, 1995–2015, but is predicted to decrease in 2025, and this is in line with the stable transition predicted between 2015 and 2025.

Table 7.4 LULC change dynamic for the period 1995–2015 and the change dynamic for the simulation and prediction using ANN-CA model for 2025

					Δ 1995–2005		Δ 2005–2015		Δ 2015–2025	
LULC classes	1995 (ha)	2005 (ha)	2015 (ha)	2025 (ha)	(ha)	%	(ha)	%	(ha)	%
Forest	633,209	438,105	267,204	222,762	366,005	44.87	170,901	20.95	44,442	5.45
Oil palm	15,524	114,066	260,410	281,067	−98,542	−12.08	−146,344	−17.94	−20,657	−2.53
Open area	93,501	50,216	19,613	8742	73,888	90.06	30,603	3.75	10,871	1.33
Paddy	15,471	16,955	10,722	9845	4749	00.58	6233	0.76	877	0.11
Urban	48,855	178,268	235,303	272,077	−186,448	−22.85	−57,035	−6.99	−36,774	−4.51
Water bodies	9229	18,179	22,537	21,297	−13,308	−1.63	−4358	−0.53	1240	0.15

Note: Area (ha) was calculated based on raster cell measured at 30 × 30 m from Landsat 8 image data processing; supervised classification of MLC algorithm

Table 7.5 Transition matrix for 1995–2005

Year	2005						
	LULC classes	Forest	Oil palm	Open area	Paddy	Urban	Water bodies
1995	Forest	0.644	0.147	0.051	0.012	0.132	0.014
	Oil palm	0.333	0.350	0.050	0.007	0.230	0.029
	Open area	0.131	0.071	0.136	0.020	0.621	0.021
	Paddy	0.220	0.184	0.052	0.465	0.068	0.010
	Urban	0.175	0.119	0.057	0.005	0.595	0.048
	Water bodies	0.122	0.043	0.053	0.004	0.335	0.442

Table 7.6 Transition matrix for 2005–2015

Year	2015						
	LULC classes	Forest	Oil palm	Open area	Paddy	Urban	Water bodies
2005	Forest	0.541	0.327	0.011	0.002	0.109	0.009
	Oil palm	0.120	0.606	0.012	0.010	0.235	0.017
	Open area	0.097	0.320	0.103	0.003	0.428	0.039
	Paddy	0.061	0.227	0.092	0.472	0.070	0.077
	Urban	0.050	0.143	0.030	0.002	0.736	0.039
	Water bodies	0.085	0.136	0.055	0.003	0.372	0.350

Table 7.7 Transition matrix for 2015–2025

Year	2025						
	LULC classes	Forest	Oil palm	Open area	Paddy	Urban	Water bodies
2015	Forest	0.787	0.204	0.001	0.000	0.008	0.001
	Oil palm	0.030	0.843	0.001	0.000	0.125	0.000
	Open area	0.035	0.052	0.409	0.000	0.258	0.046
	Paddy	0.003	0.074	0.000	0.918	0.002	0.003
	Urban	0.015	0.020	0.001	0.000	0.954	0.010
	Water bodies	0.015	0.030	0.000	0.000	0.175	0.779

7.4 Discussion

In the process of data acquisition, all the images selected are those with the lowest cloud cover percentage, less than 10%, as suggested by Hua and Ping (2018). Overall, most parts of the forested land are found in the northern region and some in the eastern region. The area is likely to be less affected by the land-use changes due to its topography being elevated in the east and forming part of the main range of Peninsular Malaysia. Furthermore, some of the forested lands has already been gazetted for conservation purpose: the peat swamp forest in the northern region comprising Raja Musa Forest Reserve (36,938 ha), Sungai Karang Forest Reserve (36,654 ha) and Sungai Dusun Wildlife Reserve (4891 ha) (Adila et al., 2017). This is among the most prominent remaining peat swamp forests in Peninsular Malaysia.

In Peninsular Malaysia, agricultural growth and extension of permanently planted land have always been the primary cause of deforestation or forest conversion. The country was once a primarily resource-based economy, with major exporting commodities consisting of minerals, timber and other tree crops such as rubber and palm oil. In the early 1960s, rapid development in the timber industries necessitated the employment of efficient tools and equipment, such as chainsaws and heavy machinery for timber harvesting and production, which hastened and exacerbated the deforestation of forest Peninsular Malaysia (Norizah et al., 2011). During that time, timber export was the major source of foreign exchange earnings for Malaysia. In the 1970s, besides timber production, Malaysia experienced another round of extensive deforestation before it slowed down substantially after the 1980s. The cause of deforestation in this period was mainly due to the expansion of agricultural land, especially for oil palm plantations (Miyamoto et al., 2014; Shevade & Loboda, 2019). This trend of declining forested areas has continued until now, and urban expansion is the main culprit contributing to the continual loss of forest areas (Wan Mohd Jaafar et al., 2020; Hua & Ping, 2018; Kanniah & Ho, 2017). Collectively, the human factor, such as the growing population, and human-related activities such as rapid industrialisation and urbanisation with social and economic changes play significant roles in raising the demand pressure for the forest area.

Over the years, demand for more agricultural land has been one of the major drivers of deforestation and forest conversion in this country, especially large tracts suitable for oil palm plantations (Shevade & Loboda, 2019). According to Varkkey et al. (2018), the expansion rate of oil palm plantations during the 1990s–2000s was doubled during this period; however, it slowed down between 2003 and 2013 due to a pledge by the authority to keep 50% of forest cover intact. Nevertheless, agricultural lands characterised by oil palms dotted the maps of 2005 and 2015, and the aggregate acreage continues to increase as shown in the predicted map of 2025. Xu et al. (2020) and Shaharum et al. (2020) stressed that the oil palm plantation has been one of the major lands used in Malaysia. Currently, Malaysia accounts for 34.3% of the total palm oil trade in the global market (Parveez et al., 2021). The high chances for this commodity crop to expand further are inevitable due to the earnings it brings. Another agricultural crop, paddy, is found concentrated in the northern region of Selangor near to the peat swamp forest in the vicinity. Paddy cultivation is another source of income for more than 40% of rice farmers in Peninsular Malaysia, particularly the indigenous peoples, and it is the third most dominant agricultural food product (Firdaus et al., 2020).

Besides, urbanisation and industrialisation are other prominent factors of the forest cover changes happening around Peninsular Malaysia. As studied by Hua and Ping (2018), Kuala Lumpur experienced high temperature in a highly built-up area and consistent decrement of the green and forest areas. The rate of urbanisation is in line with the 10th and 11th Malaysia Plans to transform the rural areas into more liveable environments with better social and economic facilities. This will improve the well-being of the rural communities; at the same time, the upgrading of infrastructures and new construction developments in the urban areas will raise the living quality and commercial competitiveness of the existing cities (Hasan & Nair,

2014; van Grunsven & Benson, 2020). These arguments are justified by other previous studies on the forest cover losses in various parts of Malaysia for urbanisation purposes (Wan Mohd Jaafar et al., 2020; Kanniah & Ho, 2017; Kanniah et al., 2015). Urban and open areas are relatively common in the central parts of Selangor, extending towards the southern region and connected to Kuala Lumpur City. Throughout the 30 years spanning 1995–2025, urban developments have resulted in a decrease in vegetation areas in the central and southern regions of Selangor and Kuala Lumpur. This scenario has also been predicted by Noor et al. (2013) by using images from the 1990–2010 periods. The population growth and the increase in socio-economic activities in this area are some factors driving the urban expansion. If the current population growth remains the same in the next 3 years, the predicted urban area expansion in 2025 will become a reality. This reasoning and prognosis are supported by other works focusing on urban research, such as that of Egidi et al. (2020) and Salvati and Lamonica (2020).

Meanwhile, waterbodies comprise mostly rivers and lakes which can be found throughout the study area, some of which act as storage reservoirs for the surrounding areas. Waterbodies are detected to increase from 1995 to 2025, and this might be due to the detection of pixels of the waterbodies' spectral signature when some areas are opened subsequently, which could happen when forest cover canopy disappears due to agricultural development and urban expansion.

7.5 Conclusion

The use of ANN-CA modelling in this chapter shows the transitions, patterns and future simulations of LULC changes based on LULC over two decades and two spatial variables, namely, DEM and distance to the road. Although the classifications of LULC for previous years are validated based on GE, the accuracies are acceptable with Cohen's kappa values of 0.81, 0.76 and 0.73, respectively, for images of 1995, 2005 and 2015.

Based on the satisfactory results in comparing the actual and predicted maps of 2015, the simulations of predicting LULC in 2025 are conducted. The ANN-CA model reveals the transitions between the initial year and final year of the 10-year intervals of 1995–2005 and 2005–2015. The findings are as follows: the forest areas, comprising the natural ecosystem and richness of biodiversity, are facing enormous pressure as a result of LULC changes and are expected to face further pressure in the run-up to 2025 due to oil palm plantation and urban area expansion. The growth of population entails further building developments in the urban areas to accommodate more people, while the agricultural land acreage needs to be increased to provide food and income. These basic human needs are the common cycles in the LULC changes and have been proven in the simulations of the transition matrix in this chapter. The consequences of these LULC changes may have adverse effects on the population and hence require careful management by the managers and the responsible stakeholders to promote sustainable urbanisation and agricultural development so that the remaining forest areas can be conserved for future generations. In future studies, other explanatory variables are recommended in running simulations of LULC with the ANN-CA model to explore the effects on the patterns of LULC changes.

Appendix 1: Validation Point Distribution with Global Moran's Index

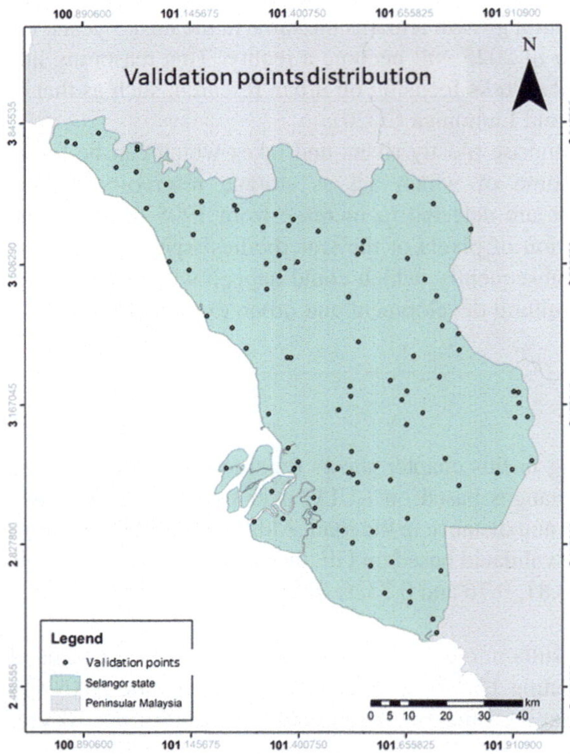

Moran's Index: 0.049326

z-score: 0.619151

p-value: 0.535817

Appendix 2: (a) Digital Elevation Model (DEM), and (b) Road Network in the Study Area

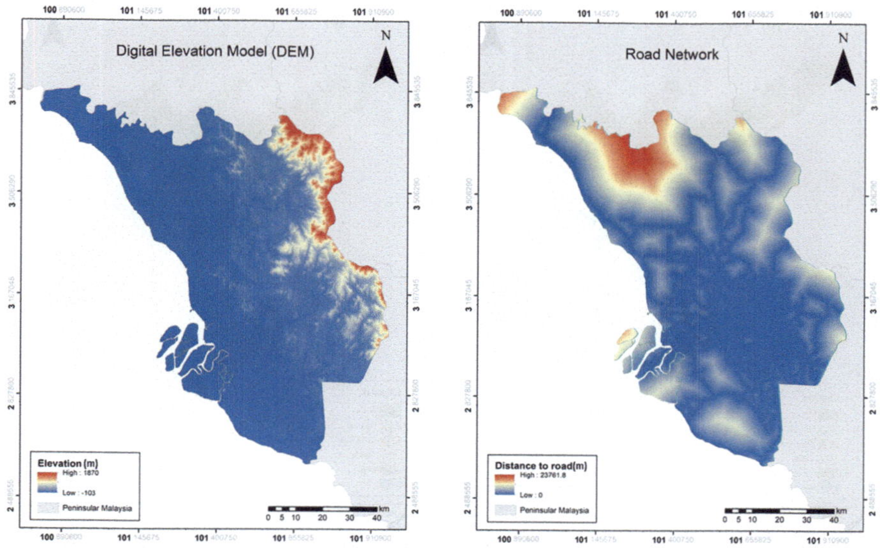

Appendix 3: (a) Real Map, and (b) Predicted Map for 2015

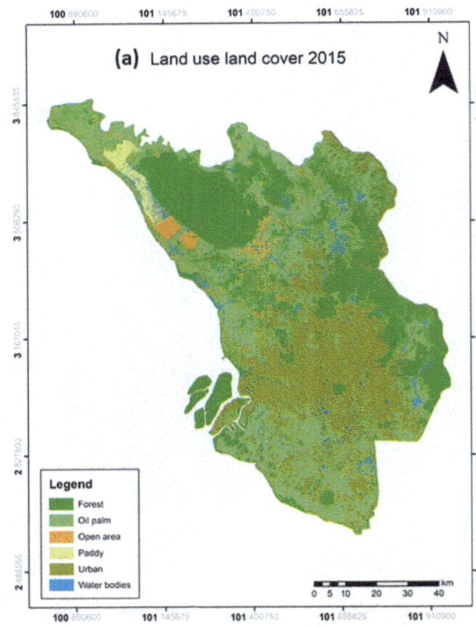

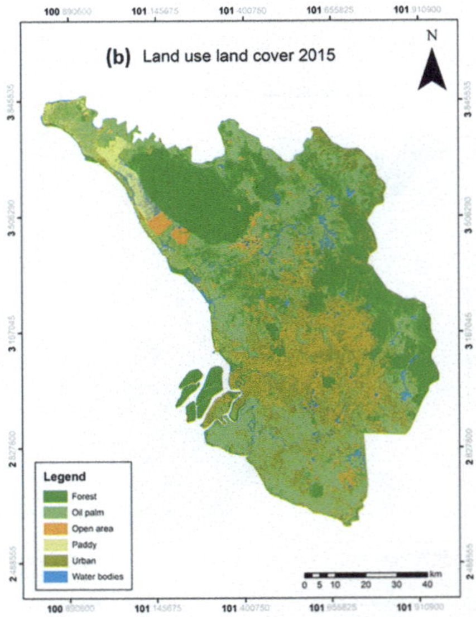

(a) LULC map for 2015 derived from MLC algorithm, (b) Predicted LULC map for 2015 derived from LULC map of 1995 and 2005.

Appendix 4: Validation Graph and Kappa Statistics

% of Correctness: 82.85880
Kappa (overall): 0.75406
Kappa (Histo): 0.85110
Kappa (loc): 0.88599

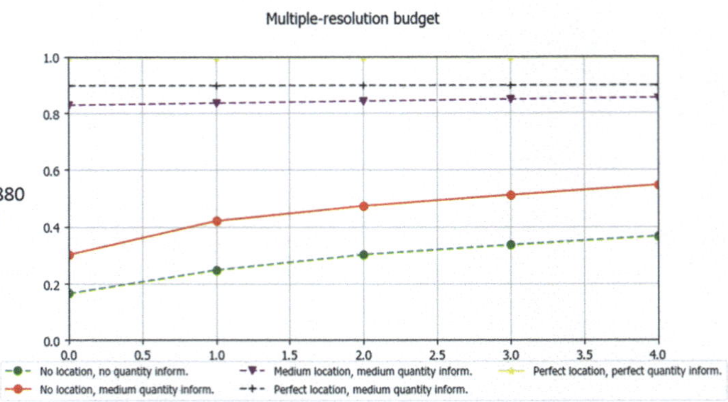

Validation graph between real LULC for 2015 and the predicted LULC 2015

References

Abbas, Z., Yang, G., Zhong, Y., & Zhao, Y. (2021). Spatiotemporal change analysis and future scenario of LULC using the CA-ANN approach: A case study of the greater bay area, China. *Land, 10*(6), 584.

Abdul-Hamid, A. Q., Ali, M. H., Tseng, M. L., Lan, S., & Kumar, M. (2020). Impeding challenges on industry 4.0 in circular economy: Palm oil industry in Malaysia. *Computers & Operations Research, 123*, 105052.

Abdullah, S. A., & Nakagoshi, N. (2008). Landscape ecological approach in oil palm land use planning and management for forest conservation in Malaysia. In *Landscape ecological applications in man-influenced areas* (pp. 179–191). Springer.

Aburas, M. M., Abdullah, S. H., Ramli, M. F., & Ash'aari, Z. H. (2015). Measuring land cover change in Seremban, Malaysia using NDVI index. *Procedia Environmental Sciences, 30*, 238–243.

Adila, N., Sasidhran, S., Kamarudin, N., Puan, C. L., Azhar, B., & Lindenmayer, D. B. (2017). Effects of peat swamp logging and agricultural expansion on species richness of native mammals in Peninsular Malaysia. *Basic and Applied Ecology, 22*, 1–10.

Aida, N., Sasidhran, S., Kamarudin, N., Aziz, N., Puan, C. L., & Azhar, B. (2016). Woody trees, green space and park size improve avian biodiversity in urban landscapes of Peninsular Malaysia. *Ecological Indicators, 69*, 176–183.

Asnawi, N. H., Ahmad, P., Choy, L. K., Syahir, M., & Khair, A. A. (2018). Land use and land cover change in Kuala Lumpur using remote sensing and geographic information system approach. *Journal of Built Environment Technology and Engineering, 4*(10), 206–216.

Atiqah, A., Rhyma, P., Jamhuri, J., Zulfa, A., Samsinar, M., & Norizah, K. (2020). Using Google Earth imagery to detect distribution of forest cover change—Is the technique practical for Malaysian forests? *The Malaysian Forester, 83*(1), 1–15.

Azhar, B., Saadun, N., Puan, C. L., Kamarudin, N., Aziz, N., Nurhidayu, S., & Fischer, J. (2015). Promoting landscape heterogeneity to improve the biodiversity benefits of certified palm oil production: Evidence from Peninsular Malaysia. *Global Ecology and Conservation, 3*, 553–561.

Beyer, E., Hagemann, A., & Misselwitz, P. (2020). Commodity flows and urban spaces. An introduction. *Articulo-Journal of Urban Research, 21*, 4522. https://doi.org/10.4000/articulo.4522

Cohen, J. (1960). A coefficient of agreement for nominal scales. *Educational and Psychological Measurement, 20*(1), 37–46.

Cumming, G. S., Buerkert, A., Hoffmann, E. M., Schlecht, E., von Cramon-Taubadel, S., & Tscharntke, T. (2014). Implications of agricultural transitions and urbanization for ecosystem services. *Nature, 515*(7525), 50–57.

Egidi, G., Halbac-Cotoara-Zamfir, R., Cividino, S., Quaranta, G., Salvati, L., & Colantoni, A. (2020). Rural in town: Traditional agriculture, population trends, and long-term urban expansion in metropolitan Rome. *Land, 9*(2), 53.

Firdaus, R. R., Leong Tan, M., Rahmat, S. R., & Senevi Gunaratne, M. (2020). Paddy, rice and food security in Malaysia: A review of climate change impacts. *Cogent Social Sciences, 6*(1), 1818373.

Foster, W. A., Snaddon, J. L., Turner, E. C., Fayle, T. M., Cockerill, T. D., Ellwood, M. F., et al. (2011). Establishing the evidence base for maintaining biodiversity and ecosystem function in the oil palm landscapes of South East Asia. *Philosophical Transactions of the Royal Society B: Biological Sciences, 366*(1582), 3277–3291.

Hasan, A. R., & Nair, P. L. (2014). Urbanisation and growth of metropolitan centres in Malaysia. *Malaysian Journal of Economic Studies, 51*(1), 87–101.

How Jin Aik, D., Ismail, M. H., & Muharam, F. M. (2020). Land use/land cover changes and the relationship with land surface temperature using Landsat and MODIS imageries in Cameron Highlands, Malaysia. *Land, 9*(10), 372.

Hua, A. K. (2017). Application of Ca-Markov model and land use/land cover changes in Malacca River Watershed, Malaysia. *Applied Ecology and Environmental Research, 15*(4), 605–622.

Hua, A. K., & Ping, O. W. (2018). The influence of land-use/land-cover changes on land surface temperature: A case study of Kuala Lumpur metropolitan city. *European Journal of Remote Sensing, 51*(1), 1049–1069.

Kanniah, K. D., & Ho, C. S. (2017). Urban forest cover change and sustainability of Malaysian cities. *Chemical Engineering Transactions, 56*, 673–678.

Kanniah, K. D., Sheikhi, A., Cracknell, A. P., Goh, H. C., Tan, K. P., Ho, C. S., & Rasli, F. N. (2015). Satellite images for monitoring mangrove cover changes in a fast growing economic region in southern Peninsular Malaysia. *Remote Sensing, 7*(11), 14360–14385.

Li, M., Zang, S., Zhang, B., Li, S., & Wu, C. (2014). A review of remote sensing image classification techniques: The role of spatio-contextual information. *European Journal of Remote Sensing, 47*(1), 389–411.

Mahamud, M. A., Samat, N., Tan, M. L., Chan, N. W., & Tew, Y. L. (2019). Prediction of future land use land cover changes of Kelantan, Malaysia. *International Archives of the Photogrammetry, Remote Sensing & Spatial Information Sciences, 2019*, 379.

Mallick, J., AlQadhi, S., Talukdar, S., Pradhan, B., Bindajam, A. A., Islam, A. R. M., & Dajam, A. S. (2021). A novel technique for modeling ecosystem health condition: A case study in Saudi Arabia. *Remote Sensing, 13*(13), 2632.

Marzuki, A., & Jais, A. S. (2020). Urbanisation and the concerns for food security in Malaysia. *Planning Malaysia, 18*(13), 202–217.

Miyamoto, M., Parid, M. M., Aini, Z. N., & Michinaka, T. (2014). Proximate and underlying causes of forest cover change in Peninsular Malaysia. *Forest Policy and Economics, 44*, 18–25.

Noor, N. M., Abdullah, A., & Manzahani, M. N. H. (2013). Land cover change detection analysis on urban green area loss using GIS and remote sensing techniques. *Planning Malaysia, 11*(3), 111.

Norizah, K. M., Hasmadi, I., Jusoff, K., & Sood, A. M. (2011). Evolution and development of forest harvesting in Peninsular Malaysia. *The Malaysian Forester, 74*(2), 79–102.

Parveez, G. K. A., Tarmizi, A. H. A., Sundram, S., Loh, S. K., Ong-Abdullah, M. E. I. L. I. N. A., Palam, K. D. P., et al. (2021). Oil palm economic performance in Malaysia and R&D progress in 2020. *Journal of Oil Palm Research, 33*(2), 111.

Reza, M. I. H. (2014). Measuring forest fragmentation in the protected area system of a rapidly developing Southeast Asian tropical region. *Science Postprint, 1*(1), e00030. https://doi.org/10.14340/spp.2014.09a0001

Salvati, L., & Lamonica, G. R. (2020). Containing urban expansion: Densification vs greenfield development, socio-demographic transformations and the economic crisis in a Southern European City, 2006–2015. *Ecological Indicators, 110*(105), 923.

Saputra, M. H., & Lee, H. S. (2019). Prediction of land use and land cover changes for north Sumatra, Indonesia, using an artificial-neural-network-based cellular automaton. *Sustainability, 11*(11), 3024.

Sasidhran, S., Adila, N., Hamdan, M. S., Samantha, L. D., Aziz, N., Kamarudin, N., et al. (2016). Habitat occupancy patterns and activity rate of native mammals in tropical fragmented peat swamp reserves in Peninsular Malaysia. *Forest Ecology and Management, 363*, 140–148.

Shaharum, N. S. N., Shafri, H. Z. M., Ghani, W. A. W. A. K., Samsatli, S., Al-Habshi, M. M. A., & Yusuf, B. (2020). Oil palm mapping over Peninsular Malaysia using Google Earth Engine and machine learning algorithms. *Remote Sensing Applications: Society and Environment, 17*, 100287.

Sheikhi, A., & Kanniah, K. D. (2018). Impact of land cover change on urban surface temperature in Iskandar Malaysia. *Chemical Engineering Transactions, 63*, 25–30.

Shevade, V. S., & Loboda, T. V. (2019). Oil palm plantations in Peninsular Malaysia: Determinants and constraints on expansion. *PLoS One, 14*(2), e0210628.

Tee, S. L., Samantha, L. D., Kamarudin, N., Akbar, Z., Lechner, A. M., Ashton-Butt, A., & Azhar, B. (2018). Urban forest fragmentation impoverishes native mammalian biodiversity in the tropics. *Ecology and Evolution, 8*(24), 12506–12521.

Teuscher, M., Gérard, A., Brose, U., Buchori, D., Clough, Y., Ehbrecht, M., et al. (2016). Experimental biodiversity enrichment in oil-palm-dominated landscapes in Indonesia. *Frontiers in Plant Science, 7*, 1538.

Thomas, N., Lucas, R., Bunting, P., Hardy, A., Rosenqvist, A., & Simard, M. (2017). Distribution and drivers of global mangrove forest change, 1996–2010. *PLoS One, 12*(6), e0179302.

van Grunsven, L., & Benson, M. (2020). *Urban development in Malaysia: Towards a new systems paradigm*. Retrieved July 28, 2021, from https://thinkcityinstitute.org/wp-content/uploads/2020/11/Issue-2.pdf

Varkkey, H., Tyson, A., & Choiruzzad, S. A. B. (2018). Palm oil intensification and expansion in Indonesia and Malaysia: Environmental and socio-political factors influencing policy. *Forest Policy and Economics, 92*, 148–159.

Wan Mohd Jaafar, W. S., Abdul Maulud, K. N., Muhmad Kamarulzaman, A. M., Raihan, A., Md Sah, S., Ahmad, A., Saad, S. N., Mohd Azmi, A. T., Jusoh Syukri, N. K., & Razzaq Khan, W. (2020). The influence of deforestation on land surface temperature—A case study of Perak and Kedah, Malaysia. *Forests, 11*(6), 670.

Xu, Y., Yu, L., Li, W., Ciais, P., Cheng, Y., & Gong, P. (2020). Annual oil palm plantation maps in Malaysia and Indonesia from 2001 to 2016. *Earth System Science Data, 12*(2), 847–867.

Zulfa, A. W., Norizah, K., Hamdan, O., Zulkifly, S., Faridah-Hanum, I., & Rhyma, P. P. (2020). Discriminating trees species from the relationship between spectral reflectance and chlorophyll contents of mangrove forest in Malaysia. *Ecological Indicators, 111*, 106024.

Linking Blue-Green Infrastructure to Microclimate and Human Thermal Comfort for Urban Cooling: A Review

8

Ruzana Sanusi

Abstract

In cities, urban heat island (UHI) effect is one of the most distinct climate issues where it is characterized by warmer conditions in city centres than the surrounding rural areas. This may influence the urban residents, especially on the health hazard issues such as thermal stress. Blue-Green Infrastructure is a nature-based solution that can improve the microclimate conditions of urban environments and improve human thermal comfort. However, the nature, imbalances and gaps of research on Blue-Green Infrastructure should be reviewed to see the extent of research on how its utilization can be linked to microclimate and human thermal comfort. Therefore, this review addresses this issue based on the previous literature. From 2018 to 2021, there was a gradual increase in the Blue-Green Infrastructure research. More research focused on Green Infrastructure followed by the combination of Blue-Green Infrastructure; however, limited studies were on Blue Infrastructure. Most studies used biophysical modelling followed by fieldwork, and only 36% studied human thermal comfort at micro level. This suggests that more studies should be done in the field to link the contribution of Blue-Green Infrastructure to urban microclimate and consequently its effect on thermal stress of urban residents. The review is concluded by highlighting aspects of Blue-Green Infrastructure research that can be further studied to improve future urban planning efforts for urban cooling.

R. Sanusi (✉)
Institute of Tropical Forestry and Forest Products (INTROP), Universiti Putra Malaysia (UPM), Serdang, Selangor, Malaysia

Faculty of Forestry and Environment, Universiti Putra Malaysia (UPM), Serdang, Selangor, Malaysia
e-mail: ruzanasanusi@upm.edu.my

Z. Samdin et al. (eds.), *Tropical Forest Ecosystem Services in Improving Livelihoods For Local Communities*, https://doi.org/10.1007/978-981-19-3342-4_8

Keywords

Nature-based solution · Green spaces · Blue spaces · Urban climate · Urban forestry · Thermal stress

8.1 Introduction

Urbanization has changed the land cover of an area. From the natural environment, the area is being converted to a built environment. The changes to the urban built environment have altered the environment due to the significant change of the built area materials, substituted with artificial materials such as concrete and asphalt. Combining the artificial materials with sealed surfaces and less vegetation will influence the radiation balances of the urban environment (Butera et al., 2018).

Urban population is one of the main drivers of urbanization where more than half of the global population are currently residing in the urban areas, and it is projected that in 2050, this will increase up to 68% (United Nations, 2018). With the population growth, many more developments will be made available, and this is unavoidable to meet the needs of urban residents. This will further deteriorate the urban environment and, specifically, will alter the heat energy balances and consequently the urban climate (Mohajerani et al., 2017). The urban climate is typically characterized by higher air temperature and drier conditions compared to the surrounding rural area, and this is often called as urban heat island (UHI) effect (Zhao et al., 2014). The increase in temperature in cities exacerbated by urban heat and global warming can reduce the thermal comfort of the urban residents.

One of the strategies in mitigating the adverse effect of urban climate issues is the implementation of Blue-Green Infrastructure that promotes urban cooling through the nature-based solution (Sanusi & Bidin, 2020). Due to its ecosystem services, Blue-Green Infrastructure has been utilized in the urban environment as part of the planning strategy. However, with the detrimental effect of landscape changes that influences microclimate conditions and human thermal comfort, there is still a lack of information on how the utilization of the Blue-Green Infrastructure in the urban area can improve urban cooling and thus humans. From a systematic review by Lourdes et al. (2021), they found that recreation, mental and physical health ($n = 54$) and moderation of extreme events services ($n = 54$) were the most studied ecosystem services among the provisioning, regulating, supporting and cultural domains. This shows that the influence of Blue-Green Infrastructure on urban climate and human well-being is an important issue, and therefore, studies that relate Blue-Green Infrastructure to microclimate and human thermal comfort should be emphasized in future urban studies. This paper aimed to review the past literature covering the current status of research on Blue-Green Infrastructure on microclimate and human thermal comfort. It addresses three specific research questions as follows:

1. Does the Blue-Green Infrastructure on microclimate and human thermal comfort research focus on specific types of Blue and Green Infrastructure or the

combination of Blue-Green Infrastructure? In what year and country they are being assessed?
2. What are the data collection processes of the Blue-Green Infrastructure research on microclimate and human thermal comfort and what types of Blue-Green structures are being assessed?
3. Does the research done on Blue-Green Infrastructure specifically measure human thermal comfort?

Further recommendations on the future research direction based on the findings will be given to conclude this review.

8.2 Materials and Method

In this review, literature from previous studies was searched using search string composed terms of the selected topic in the Scopus database. The search string composed terms were focused on the specific research concerning Blue-Green Infrastructure, microclimate and human thermal comfort. The search string applied to the database is as shown below:

("blue-green infrastructure" OR "green space" OR "green infrastructure" AND "blue spaces" OR "blue infrastructure") AND "microclimate" AND "human thermal comfort" OR "thermal stress" AND "urban" OR "city" OR "cities".

From the literature search string on Blue-Green Infrastructure, microclimate and human thermal comfort, a total of 61 articles were returned including journal articles, reviews and book chapters (Fig. 8.1). For the document type, 17 reviews and a book chapter were excluded as this study includes only journal articles. Moreover, 12 papers were further excluded as there was no clear relation to the investigated topics. In these performed searches, all abstracts were read carefully to extract important information that can provide some insight into the research within the context of the selected topic. Specifically, this paper looked into six types of information which were (1) types of infrastructure, (2) year of studies, (3) country of origin, (4) data collection process, (5) Blue-Green structure and (6) human thermal comfort studies (Fig. 8.1).

8.3 Results and Discussion

8.3.1 General Patterns of Returned Articles

From the searches, the trend of research on Blue-Green Infrastructure to microclimate and human thermal comfort was acquired, where only 33 papers were related to this topic. This shows that future studies should be further expanded on the climate and human thermal comfort perspectives as in the context of urban areas, urban heat

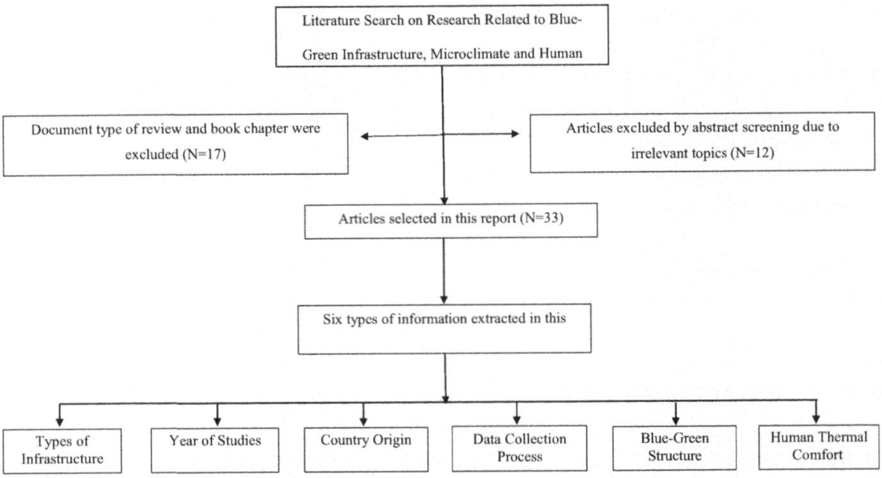

Fig. 8.1 Schematic representation from the search string on articles related to Blue-Green Infrastructure, microclimate and human thermal comfort

island poses a threat to heat-related human health issues (Aghamohammadi et al., 2021). On the other hand, the research trend or direction towards the impact of urban Blue-Green Infrastructure on urban climate and human well-being in Southeast Asia is also quite promising (Lourdes et al., 2021).

From a thorough analysis of the returned articles, not all the research focused on the specific combination of Blue-Green Infrastructure (Fig. 8.2). Only 33% of the returned articles researched the contribution of Blue-Green Infrastructure on urban cooling ($n = 11$). On the other hand, green infrastructure was the main focus in the research with 55% ($n = 18$), while the lowest studied was blue infrastructure with only 12% ($n = 4$). Although research that combines blue and green infrastructures is still progressing, it was unbalanced and needed further studies to see its contribution to urban climate and human thermal comfort especially the blue infrastructure. This is because Blue-Green infrastructure is seen to be a future key strategy in sustainably planning and managing the urban environment (Din Dar et al., 2021). For instance, applying both elements in parks could reduce the UHI due to its synergistic cooling capability (Gunawardena et al., 2017). However, although the application of Blue-Green Infrastructure has received great interest globally due to its ecosystem services (Dai et al., 2021; Nguyen et al., 2021), there is still a need to further determine the influence of applying the Blue-Green Infrastructure in urban environments (Ghofrani et al., 2017). Lack of understanding on how the Blue-Green Infrastructure can be beneficial and in what way it is best to be implemented in urban environments may also lead to future exploration of its benefits and application (Bedla & Halecki, 2021).

Moreover, from these findings, it was apparent that the green infrastructure was predominantly studied. This might be due to green infrastructures being widely used in mitigation strategies in combating urban climate issues such as the UHI. The

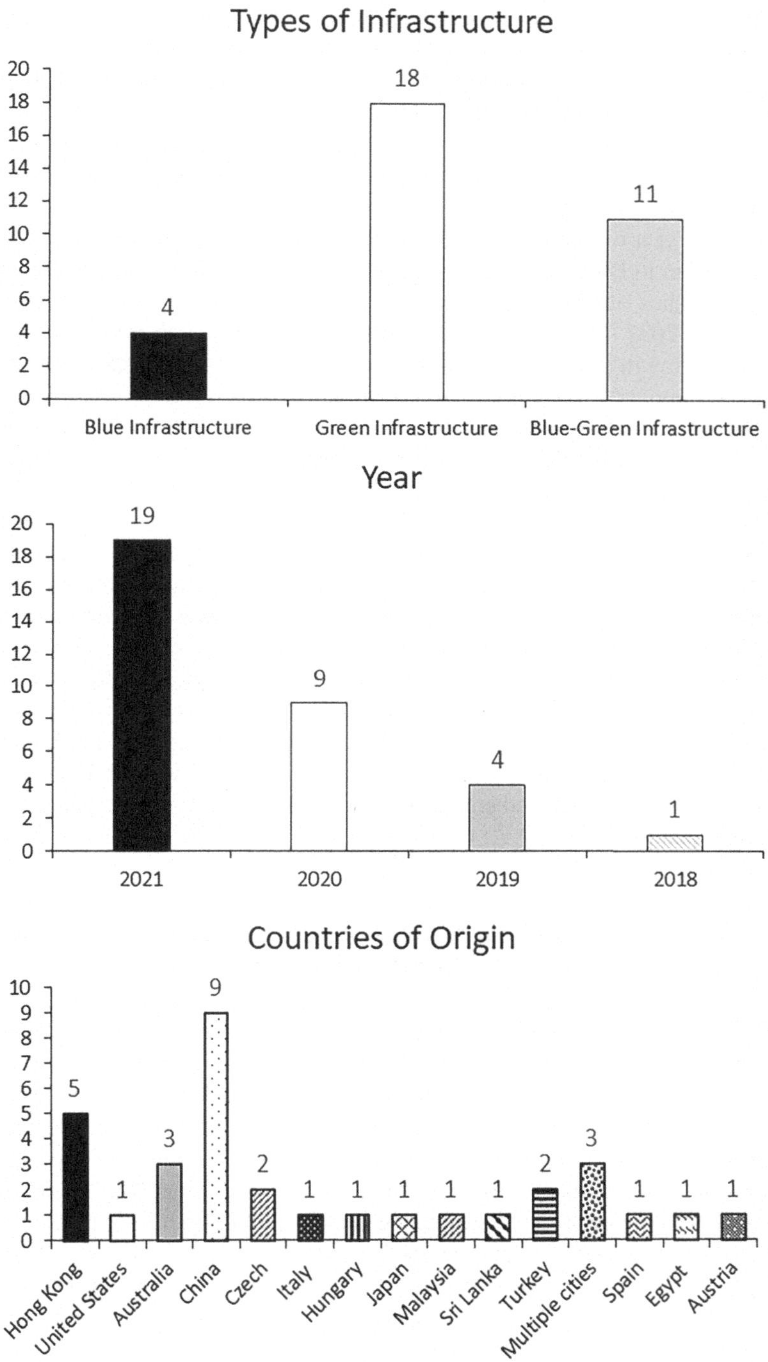

Fig. 8.2 The general information extracted from all the returned articles of types of infrastructure, year of publication and countries of research origin

green infrastructure not only provides various environmental benefits including the urban cooling and mitigation of UHI but also consequently serves the urban residents with a thermally comfortable environment, thus improving their health and well-being (Wilis & Petrokofsky, 2017). Furthermore, the possibility of the green and blue spaces that could differently affect the cooling effect of the environment (Hu & Li, 2020) may also lead to more studies being done to look at the infrastructure individually.

In terms of year of publication, when relating the urban microclimate and human thermal comfort to Blue-Green Infrastructure, it was noticeable that relevant articles from the searches only started in the year 2018 (3%) and a gradual increase was apparent until 2021 (58%) (Fig. 8.2). Thus the spike of research from 2018 to 2021 signified the growth of interest in linking the Blue-Green Infrastructure contribution to urban cooling and human health and well-being.

In addition, despite the growth of interest in this research area, it was largely studied in China with 28%, followed by Hong Kong with 15%, Australia and multiple cities (study conducted in more than one city) with 9% and the Czech Republic and Turkey with 6% (Fig. 8.2). Moreover, countries such as Italy, Hungary, Japan, Malaysia, Sri Lanka, Spain, Egypt and Austria were with 3% each. According to de Macedo et al. (2021), cities in China also had a high contribution of studies on the concept of urban Blue-Green Infrastructure in the Global South, specifically related to local sustainable development. This indicates that the interests of big cities like those in China are to strategically address the urban issues through the application of Blue-Green Infrastructure. It is also suggested that most studies on the ecosystem services, at least in Southeast Asia, are predominantly carried out in more developed countries such as Singapore, Thailand, Indonesia and Malaysia, while less developed countries such as Myanmar, Cambodia and Laos are less studied (Lourdes et al., 2021).

8.3.2 Data Collection Process, Blue-Green Infrastructure Structure and Human Thermal Comfort Studies

For all studies, their data collection process varies according to the objectives they wanted to achieve. All the data collection involved the microclimate measurements; however, all the measurements differed depending on the data collection approach. From the returned articles, four types of data collection involved biophysical modelling, field study, mixed-method and survey (Table 8.1, Fig. 8.3). In terms of microclimate measurements for the biophysical modelling, the microclimate parameters were typically collected using on-site mobile or fixed meteorological stations and remote sensing data, where these data were calculated, simulated and modelled according to the desired urban landscapes and their conditions for each study (Table 8.1). Similarly, the mixed method also involved the mobile or fixed meteorological stations and remote sensing data but mixed with other data collections (Table 8.1). For mixed method, it involved the combination of different approaches as follows: (1) biophysical modelling and fieldwork, (2) biophysical

Table 8.1 The main findings of the returned articles related to Blue-Green Infrastructure, microclimate and human thermal comfort

No.	Authors	Location	BGI	Data collection process	Blue and green structure	Human thermal comfort studies	Findings
1.	Vo and Hu (2021)	NYC, US	BGI	Process/mechanistic/ land cover (biophysical modelling)	Multiple BG spaces	No	Diurnal analysis of canopy temperature to understand the urban landscapes and the surrounding ambient conditions Greenspace coverage and distance to blue spaces determined the cooling potential of tree canopy
2.	Lan et al. (2021)	Hong Kong	BGI	Process/mechanistic/ land cover (biophysical modelling)	Multiple BG spaces	No	Using mobile meteorological stations for microclimate data and simulating changes in urban landscapes and their effect on the environment through microclimate mode1ing Green infrastructure (turf and green facades) in proximity to water bodies can further the urban cooling benefits
3.	Cheung et al. (2021a)	Hong Kong	GI	Mixed approach: Field and unmanned aerial vehicle (UAV)	Golf	No	To link land surface temperature (LST) of trees and the below-canopy thermal conditions by measuring the differences and correlations between LST and thermal condition under the tree

(continued)

Table 8.1 (continued)

No.	Authors	Location	BGI	Data collection process	Blue and green structure	Human thermal comfort studies	Findings
							Background air temperature (ΔTcs only), relative humidity, mean radiant temperature and shortwave radiation significantly influenced the differences between tree canopy temperature and below-canopy ground surface temperature (ΔTcs) as well as between tree canopy temperature and below-canopy air temperature (ΔTca)
4.	Herath et al. (2021)	Melbourne	GI	Process/mechanistic/ land cover (biophysical modelling)	Multiple green spaces	No	Simulations of the urban climate to look at the heat mitigation performance of different urban surface parameters Green roofs lowered −1.15 °C of air temperature at night. Trees provided greater cooling potential when planted in both street canyons and parks compared to when only in the street

5.	Zhu et al. (2021)	Jinan, China	BGI	Process/mechanistic/ land cover (biophysical modelling)	Park	No	Examining the thermal environment of parks through LST Park beneficially cooling the area with the park cooling area was approximately 120.68 ha. Park cooling island potential is dependent on water area proportion, park perimeter and greenness. Increasing vegetation contributes to a greater park cooling area
6.	Chen et al. (2021)	Guiyang, China	GI	Process/mechanistic/ land cover (biophysical modelling)	Remnant	No	Linking the urban remnant natural mountains (URNMs) to examine the cool island effect and its influence factors for cooling through LST URNMs are the key factor in reducing land surface temperature and contributes to the surrounding cool island effects with an average cooling distance of approximately 170 m
7.	Lehnert et al. (2021)	Czech Republic: Brno, Olomouc, Ostrava and Plzeň	BGI	Field	Multiple BG spaces	Yes	On-site microclimate measurements in determining the human thermal comfort in different urban settings The greatest universal thermal climate index (UTCI) improvement was influenced by high vegetation with trees

(continued)

Table 8.1 (continued)

No.	Authors	Location	BGI	Data collection process	Blue and green structure	Human thermal comfort studies	Findings
							provided up to 10.5 °C thermal comfort benefits. However, with low vegetation, water fountains, spray fountains and misting systems, the cooling effect was insignificant or low
8.	Gatto et al. (2021)	Lecce city (southern Italy)	GI	Process/mechanistic/ land cover (biophysical modelling)	Tree	Yes	Greening, building geometry and microclimate data were measured on-site and employed in the simulation of different scenarios through Computational Fluid Dynamics-based and microclimate model ENVI-met to determine their interactions and impacts Trees reduced the daily air temperature, mean radiant temperature and predicted mean vote by up to 1.0 °C, 5.5 °C and 0.53, respectively. These reductions were more apparent with the increase in urban greening

9.	Huang et al. (2021)	Beijing, China	GI	Process/mechanistic/ land cover (biophysical modelling)	Multiple green spaces	No	Multiple data and methodology including meteorological data, satellite images, electronic maps, questionnaire survey data, statistical yearbooks and ArcGIS to determine the impact of the urban thermal environment influence the health of residents This study further proved that UHI impacted the human health of the urban residents by affecting their respiratory diseases, cardiovascular diseases and emotional health especially in the south of the North Second Ring Road in central Beijing
10.	Gál et al. (2021)	Szeged (Hungary)	GI	Process/mechanistic/ land cover (biophysical modelling)	Multiple green spaces	Yes	Parameters of air temperature and humidity, balanced heat and moisture budgets in the soil, shortwave and longwave radiation and the effect of clouds, used in the model simulations to determine the role of green spaces on heat stress modification Different types of urban green spaces lead to a different cooling potential, but overall, urban green spaces such as parks contribute to urban cooling. Furthermore, large urban parks provided the greatest reduction of heat load

(continued)

Table 8.1 (continued)

No.	Authors	Location	BGI	Data collection process	Blue and green structure	Human thermal comfort studies	Findings
11.	Cheung et al. (2021b)	Hong Kong	BGI	Field	Multiple BG spaces	No	The effect of park design and urban landscape on air temperature and relative humidity was determined through sensors deployment throughout 14 parks Only the largest park in Hong Kong had a significant mean temperature reduction of 0.6 °C. Background temperature greatly determined the cooling and humidifying potentials where parameters of proximity to the sea, shrub cover, tree cover and sky view factor were significantly affecting temperature and relative humidity. Mean temperature reduction of 0.07 °C and 0.04 °C was observed with every 10% increase of shrub and tree covers, respectively
12.	Lehnert et al. (2021)	Czech cities: Brno, Olomouc, Ostrava and Plzeň	BGI	Field	Multiple BG spaces	Yes	Biometeorological measurements were correlated with thermal sensation vote (TSV) via survey to investigate the thermal sensation TSV was low in the open

							grassy area. Heat mitigation strategies in urban planning should consider human behavioural patterns in combination with the microclimatic influence
13.	Back et al. (2021)	Tyrol, Western Austria	BGI	Process/mechanistic/ land cover (biophysical modelling)	Multiple BG spaces	Yes	Spatial modelling approach to simulate land surface temperature (LST), mean radiant temperature (MRT) and universal thermal climate index (UTCI) in a 2D environment. Human thermal comfort estimation can be affected by the sky view factor. Although high-albedo surfaces substantially reduced land surface temperature, it was observed that it would have an impact on human thermal comfort as the apparent temperature of MRT and UTCI values increased
14.	Fang et al. (2021)	Guangzhou in South China	GI	Survey	Lawn	Yes	Microclimate conditions were regressed with personal factors and human perception to see their interaction Thermal comfort was largely affected by air temperature, and with greater clothing insulation, the mean thermal sensation vote will increase

(continued)

Table 8.1 (continued)

No.	Authors	Location	BGI	Data collection process	Blue and green structure	Human thermal comfort studies	Findings
15.	Syafii (2021)	Saitama prefecture, Japan	BI	Field	Artificial pond	No	Shortwave and longwave radiations were measured to determine the effect of water bodies on urban microclimate In comparison to a flat concrete surface, waterbodies would absorb and store more heat, thus leading to lower surface temperature
16.	Teoh et al. (2021)	Ipoh, Malaysia	GI	Process/mechanistic/ land cover (biophysical modelling)	Street	Yes	GIS modelling and simulation of the urban landscape design using microclimate input to determine the thermal comfort improvement Street trees are recommended in tropical countries in improving outdoor thermal comfort due to their shading effect. However, thermal comfort was not significantly affected by wind due to site characteristics
17.	Perera et al. (2021)	Colombo, Sri Lanka	GI	Mixed approach: Process/mechanistic/ land cover (biophysical modelling) and survey	Park	Yes	Daytime temperature and humidity values to determine heat index and related to the ENVI-met model and survey Tree planting at the curb sides was observed to have a 2.07 °C temperature reduction

18.	Yilmaz et al. (2021)	Ezurum, Turkey	GI	Process/mechanistic/ land cover (biophysical modelling)	Street	Yes	Microclimate data embedded ENVI-met model to determine the outdoor thermal comfort for different landscape design scenarios Road designed with semi-open canopy provided greater thermal comfort to enable urban residents to walk and cycle in winter
19.	Xie and Li (2021)	Wuhan, China	BGI	Process/mechanistic/ land cover (biophysical modelling)	Park	No	Land surface temperature (LST) derived through Envi 5.2 based on the Landsat 8 image Majority of urban parks could offer park cool island (PCI) intensity in the range of 0.08–7.29 °C where larger and wide parks possessed stronger PCI intensity. Most significantly, PCI effect was greatest with water bodies in urban parks
20.	Santamouris and Osmond (2020)	Multiple cities	GI	Process/mechanistic/ land cover (biophysical modelling)	Multiple green spaces	No	Studies involved multiple cities using different scenarios and case studies exploring the influence of additional GI on urban temperature, air pollution and health. Although with maximum green infrastructure fraction, reduction in the average

(continued)

Table 8.1 (continued)

No.	Authors	Location	BGI	Data collection process	Blue and green structure	Human thermal comfort studies	Findings
							maximum peak daily and night-time temperature may not exceed 1.8 °C and 2.3 °C, respectively. The decrease of peak daily temperature by 0.1 °C leads to a reduction of heat-related mortality by 3%
21.	Li et al. (2020)	Zhengzhou, China	GI	Field	Street/campus	No	Through the meteorological parameters and coverage characteristic parameters' [i.e. Canopy Density (CD), leaf area index (LAI), Photosynthetically Active Radiation (PAR), Mean Leaf Angle (MLA)] measurements, different coverage types in small green spaces significantly affect the temperature, especially at noontime. Multilayer vegetation-covered tree-shrub-grass area had the largest cooling benefits in comparison to the impervious surface. This signifies that tree cover had the greatest influence on the temperature

22.	Irmak et al. (2020)	Ezurum, Turkey	BGI	Field	Multiple BG spaces	Yes	Using the weather stations that collected the micrometeorology parameters, the thermal comfort was determined using Physiologically Equivalent Temperature (PET) Index Thermal comfort in the city was greatly balanced by the dense green areas, regardless of the summer and winter seasons
23.	Lin et al. (2020)	Shenzhen and Hong Kong and parts of two adjacent fast-growing cities (Dongguan and Huizhou)	BI	Process/mechanistic/ land cover (biophysical modelling)	Multiple blue spaces	No	Land surface temperature determined the surface urban heat island intensity (SUHI) SUHI was reduced by 11.33% due to a 10% increase in the waterbody. However, the cooling potential may be reduced with the irregular shape of the water bodies
24.	Cheung et al. (2020)	Hong Kong	GI	Field	Multiple green spaces	Yes	Long-term meteorological conditions were monitored at four peri-urban woodland sites and a rooftop control site to determine human thermal comfort Increase in daily total incoming shortwave radiation could enhance the mean cooling magnitude in air temperature, physiological equivalent temperature and universal thermal climate index by 0.03 °C, 0.16 °C and 0.08 °C, respectively

(continued)

Table 8.1 (continued)

No.	Authors	Location	BGI	Data collection process	Blue and green structure	Human thermal comfort studies	Findings
25.	Shi et al. (2020)	Chongqing, China	BGI	Mixed approach: Process/mechanistic/ land cover (biophysical modelling) and field	Multiple BG spaces	No	Air temperature and relative humidity at different land-use sites of forest, lawn and impervious pavement with and without water areas used in the ENVI-met simulation to look at Synergistic Cooling Effects (SCEs) of green-blue space Green-blue spaces in the proximity of 7–12 m to waterbody could increase the SCEs, as there was a 3.3 °C greater mean air temperature reduction in this area than individual water and forest areas
26.	Meili et al. (2020)	Telok Kurau, Singapore, Preston, Melbourne, Australia and Maryvale, Phoenix, USA	GI	Process/mechanistic/ land cover (biophysical modelling)	Multiple green spaces	No	– The urban ecohydrological model of Urban Tethys-Chloris (UT & C) used meteorological time series of air temperature, humidity, air pressure, incoming shortwave and longwave radiation, precipitation and wind speed to look at the effects of the urban environment on plant well-being and performance

							– Fully grass-covered ground area and high values of leaf area index in Singapore could lead to the reduction of 1.1 °C and 0.3 °C of air temperature, respectively, thus increasing the relative humidity by 6.5% and 2.1%, respectively
27.	Chàfer et al. (2020)	Puigverd de Lleida (Spain)	GI	Field	Vertical greening	No	– Relative humidity and air temperature sensors were used to record the microclimate of pergola with different shading systems of vegetation and ropes – Green infrastructure provides lower surface temperatures and evapotranspiration than a simple pergola system, where it lowered the air temperature up to 5 °C at pedestrian level and increased relative humidity in summer
28.	Fung and Jim (2020)	Hong Kong	BGI	Field	Multiple BG spaces	Yes	– Thermal comfort at a pondside lawn, open lawn and a concrete rooftop was estimated through universal thermal climate index (UTCI) that used the microclimatic parameters

(continued)

Table 8.1 (continued)

No.	Authors	Location	BGI	Data collection process	Blue and green structure	Human thermal comfort studies	Findings
							– Compared to the open lawn, the pondside lawn had a 0.7 °C lower air temperature. Although the pondside was able to reduce the air temperature, it could not improve the human thermal comfort conditions due to the lack of shading effect from the tree
29.	Aboelata and Sodoudi (2019)	Cairo	GI	Process/mechanistic/ land cover (biophysical modelling)	Street	Yes	– Air temperature, wind speed, globe temperature and humidity were used in the Envi-met simulation study of different vegetation scenarios – Human thermal comfort was 3 K cooler with a 50% increase of trees. However, factors such as energy demand in buildings, street orientation and aspect ratio should be considered in future planning
30.	Wu et al. (2019)	Wuhan, China	BI	Process/mechanistic/ land cover (biophysical modelling)	River	No	– Land surface temperature (LST) relationship to the cooling effects of urban blue infrastructure was investigated

							– The size and shape of water bodies affect the cooling potential of urban blue infrastructure as the water surface temperature decreased with larger urban blue infrastructure size
31.	Imran et al. (2019)	Melbourne	GI	Process/mechanistic/ land cover (biophysical modelling)	Multiple green spaces	Yes	Simulations using the Weather Research and Forecasting (WRF) model and Single-Layer Urban Canopy Model (SLUCM) were done to look at the effectiveness of urban vegetation patches in reducing UHI effects Meteorological input of temperature, relative humidity and solar radiation simulated by the WRF model were used to estimate the HTC Through the increment of green fraction from 20 to 50%, the mixed forest, mixed shrublands and grasslands and mixed forest and grasslands reduced near-surface (2 m) UHI by 0.6–3.4 °C, 0.4–3.0 °C and 0.6–3.7 °C, respectively HTC did not improve with vegetated patches, but a substantial improvement was observed between the evening and early morning

(continued)

Table 8.1 (continued)

No.	Authors	Location	BGI	Data collection process	Blue and green structure	Human thermal comfort studies	Findings
32.	Stanley et al. (2019)	Salzburg, Austria	GI	Field	Tree	No	Microclimate conditions of the crown-shaded and full sun-exposed area along with tree phenology and physiognomy were measured Tree shading reduced 12.2 °C of surface temperature compared to surfaces in the sun, and tree characteristics led to different cooling effects. Tree height, trunk circumference and age greatly contributed to surface cooling
33.	Cai et al. (2018)	Chongqing, China	BI	Process/mechanistic/ land cover (biophysical modelling)	River	No	Urban land surface temperature (LST) was used to look at the cooling effects of water bodies LST could increase with the increase in the distance to the water bodies. Moreover, the cooling effect of water bodies could be reached within the distance of 1 km

BGI Blue-Green Infrastructure, *GI* Green Infrastructure, *BI* Blue Infrastructure

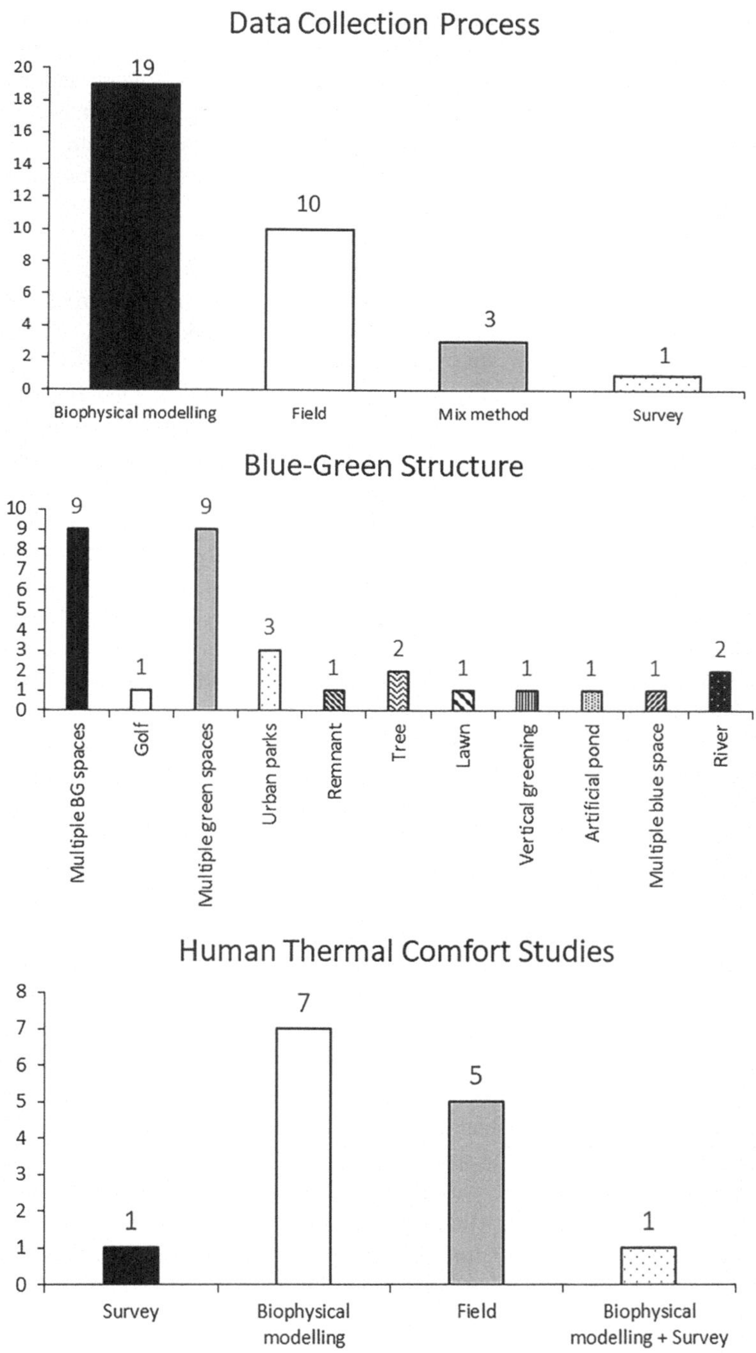

Fig. 8.3 The specific information on the data collection process, Blue-Green Infrastructure (*BGI* Blue-Green Infrastructure, *GI* Green Infrastructure, *BI* Blue Infrastructure) and human thermal comfort studies of all the returned articles

modelling and survey and (3) fieldwork and unmanned aerial vehicle. While for field studies, the microclimate parameters were measured on-site at different urban settings such as parks, water bodies, tree coverages, peri-urban woodland, rooftop, pergola and pondside (Table 8.1). On the other hand, the survey data collection involved on-site microclimate measurements that were combined with the survey questionnaire at desired urban landscapes to see the interaction of microclimate with personal factors and perception (Table 8.1).

Further analysis showed that the biophysical modelling involving the process, mechanistic or land cover modelling and analysis was mainly used as a tool in accessing the contribution of Blue-Green Infrastructure to urban microclimate and human thermal comfort with a total of 58% of the returned articles. This is then followed by field study with 30%, mixed-method with 9% and survey with 3% (Fig. 8.3). The variety of approaches in assessing the Blue-Green Infrastructure indicates that the assessment can be conducted in many ways. Finding the best methodology may be hard, but applying the suitable methodology to the context of its study is more relevant when assessing the contribution of Blue-Green Infrastructure to urban microclimate and human thermal comfort. Despite this, it is also important to note that as biophysical modelling was mainly used in the assessment, it showed that the data and findings of these studies were largely relied on the process, mechanistic or land cover modelling and analysis level rather than the site level. The biophysical modelling could help spatially plan the city and can further identify vulnerable zones through a clear and compelling model aimed to be used in the decision-making and planning processes (Khorrami & Malekmohammadi, 2021). However, field data should also be seen as one of the options as it could access the impact of Blue-Green Infrastructure at the micro level, and when talking in the context of human thermal comfort, it will be most relevant as it is directly related to the urban residents.

This review further elaborates that when researching the Blue-Green Infrastructure, it involved both blue and green elements in their research (Fig. 8.3). Most of the research involved multiple Blue-Green spaces and multiple green spaces. For green infrastructures, the structure varies where the research was done in areas of trees, golf course, urban parks, remnants, lawn and vertical greening. This further elaborates the importance of green structures such as trees in the urban microclimate and human thermal comfort as they could enhance the urban environment through urban cooling and therefore reduce thermal stress (Sanusi et al., 2016, 2017; Sanusi & Bidin, 2020). On the other hand, for blue infrastructure, it was studied in river, artificial pond and multiple blue spaces.

It is also apparent that there was alack of studies that measured human thermal comfort from all the returned articles. So far, only 14 studies were estimating the impact of Blue-Green Infrastructure on human thermal comfort, where most of the studies were using biophysical modelling, followed by fieldwork (Fig. 8.3). This indicates that most studies only relatively discussed this in general without actually estimating the impact on urban residents and the research was primarily based on a biophysical modelling approach. It has been known that many studies look at the importance of vegetation especially trees to provide ecosystem services such as mitigation of urban climate and improving human health and well-being especially for the reduction of heat-related illnesses (Chianucci et al., 2015; de Abreu-Harbich et al., 2015; Sanusi & Livesley, 2020). Therefore, without physically estimating the impact at the micro level, less evidence can be provided in estimating the human thermal comfort benefits when applying Blue-Green Infrastructure in the urban environment. This would be an important knowledge gap that is needed to address further understanding on the role of Blue-Green Infrastructure in urban areas including on how far the Blue-Green Infrastructure can ameliorate the urban climate and therefore improve the urban thermal conditions. Further description of all the extracted information from the returned articles and related findings that are discussed in this current review is detailed out and summarized in Table 8.1.

8.4 Conclusions

From this review, the Blue-Green Infrastructure research on microclimate and human thermal comfort research were specifically focused on the Green Infrastructure followed by the combination of Blue-Green Infrastructure, while studies on Blue Infrastructure were limitedly being studied upon. Moreover, there was a gradual surge in the Blue-Green Infrastructure studies with an increasing trend of 55% studies from 2018 to 2021. The findings suggest that there are still many knowledge gaps on the contribution of Blue-Green Infrastructure, and with the growing interest, there is a need to further determine the influence of applying Blue-Green Infrastructure in the urban environment, especially in terms of microclimate and human thermal comfort. Furthermore, most research was largely studied in China, followed up by Hong Kong and Australia. This further highlights the need of expanding research on related topics to many other regions.

It was also notable that biophysical modelling was the most used approach and followed by fieldwork. Moreover, only 36% of the total study measured the human thermal comfort parameters in the field, thus indicating limited proof to link Blue-Green Infrastructure's role to mitigate the urban microclimate and improve the thermal comfort of urban residents at the micro level.

The application of Blue-Green Infrastructure would synergistically improve the microclimate and the heat-related health problems for urban cooling. This review addresses the knowledge gap of Blue-Green Infrastructure research on microclimate and human thermal comfort. It is concluded that specific research should be further expanded on the discussed knowledge gaps to ensure more accurate decisions can be made for future urban planning efforts using Blue-Green Infrastructure as a nature-based solution.

Acknowledgements Appreciation goes to Universiti Putra Malaysia for the funding received from the Putra Grant-Putra Young Initiative (IPM) (Grant No: GP-IPM/2018/9637800).

References

Aboelata, A., & Sodoudi, S. (2019). Evaluating urban vegetation scenarios to mitigate urban heat island and reduce buildings' energy in dense built-up areas in Cairo. *Building and Environment, 166*, 106407. https://doi.org/10.1016/j.buildenv.2019.106407

Aghamohammadi, N., Fong, C. S., Idrus, M. H. M., Ramakreshnan, L., & Sulaiman, N. M. (2021). Environmental heat-related health symptoms among community in a tropical city. *Science of the Total Environment, 782*, 146611. https://doi.org/10.1016/j.scitotenv.2021.146611

Back, Y., Bach, P. M., Jasper-Tönnies, A., Rauch, W., & Kleidorfer, M. (2021). A rapid fine-scale approach to modelling urban bioclimatic conditions. *Science of the Total Environment, 756*, 143732. https://doi.org/10.1016/j.scitotenv.2020.143732

Bedla, D., & Halecki, W. (2021). The value of river valleys for restoring landscape features and the continuity of urban ecosystem functions–A review. *Ecological Indicators, 129*, 107871.

Butera, F., Eugenio, M., Chiara, P. M., & Fabrizio, L. (2018). *Energy and resource efficient urban neighbourhood design principles for tropical countries: A practitioner's guidebook.* UN Habitat.

Cai, Z., Han, G., & Chen, M. (2018). Do water bodies play an important role in the relationship between urban form and land surface temperature? *Sustainable Cities and Society, 39*, 487–498. https://doi.org/10.1016/j.scs.2018.02.033

Chàfer, M., Pisello, A. L., Piselli, C., & Cabeza, L. F. (2020). Greenery system for cooling down outdoor spaces: Results of an experimental study. *Sustainability (Switzerland), 12*(15), 5888. https://doi.org/10.3390/SU12155888

Chen, X., Wang, Z., & Bao, Y. (2021). Cool island effects of urban remnant natural mountains for cooling communities: A case study of Guiyang, China. *Sustainable Cities and Society, 71*, 102983. https://doi.org/10.1016/j.scs.2021.102983

Cheung, P. K., Fung, C. K. W., & Jim, C. Y. (2020). Seasonal and meteorological effects on the cooling magnitude of trees in subtropical climate. *Building and Environment, 177*, 106911. https://doi.org/10.1016/j.buildenv.2020.106911

Cheung, P. K., Jim, C. Y., & Hung, P. L. (2021a). Preliminary study on the temperature relationship at remotely-sensed tree canopy and below-canopy air and ground surface. *Building and Environment, 204*, 108169. https://doi.org/10.1016/j.buildenv.2021.108169

Cheung, P. K., Jim, C. Y., & Siu, C. T. (2021b). Effects of urban park design features on summer air temperature and humidity in compact-city milieu. *Applied Geography, 129*, 102439. https://doi.org/10.1016/j.apgeog.2021.102439

Chianucci, F., Puletti, N., Giacomello, E., Cutini, A., & Corona, P. (2015). Estimation of leaf area index in isolated trees with digital photography and its application to urban forestry. *Urban Forestry & Urban Greening, 14*(2), 377–382.

Dai, X., Wang, L., Tao, M., Huang, C., Sun, J., & Wang, S. (2021). Assessing the ecological balance between supply and demand of blue–green infrastructure. *Journal of Environmental Management, 288*, 112454.

De Abreu-Harbich, L. V., Labaki, L. C., & Matzarakis, A. (2015). Effect of tree planting design and tree species on human thermal comfort in the tropics. *Landscape and Urban Planning, 138*, 99–109.

de Macedo, L. S. V., Picavet, M. E. B., de Oliveira, J. A. P., & Shih, W. Y. (2021). Urban green and blue infrastructure: A critical analysis of research on developing countries. *Journal of Cleaner Production, 313*, 127898.

Din Dar, M. U., Shah, A. I., Bhat, S. A., Kumar, R., Huisingh, D., & Kaur, R. (2021). Blue green infrastructure as a tool for sustainable urban development. *Journal of Cleaner Production, 318*, 128474.

Fang, Z., Zheng, Z., Feng, X., Shi, D., Lin, Z., & Gao, Y. (2021). Investigation of outdoor thermal comfort prediction models in South China: A case study in Guangzhou. *Building and Environment, 188*, 107424. https://doi.org/10.1016/j.buildenv.2020.107424

Fung, C. K. W., & Jim, C. Y. (2020). Influence of blue infrastructure on lawn thermal microclimate in a subtropical green space. *Sustainable Cities and Society, 52*, 101858. https://doi.org/10.1016/j.scs.2019.101858

Gál, T., Mahó, S. I., Skarbit, N., & Unger, J. (2021). Numerical modelling for analysis of the effect of different urban green spaces on urban heat load patterns in the present and in the future. *Computers, Environment and Urban Systems, 87*, 101600. https://doi.org/10.1016/j.compenvurbsys.2021.101600

Gatto, E., Ippolito, F., Rispoli, G., Carlo, O. S., Santiago, J. L., Aarrevaara, E., et al. (2021). Analysis of urban greening scenarios for improving outdoor thermal comfort in neighbourhoods of Lecce (Southern Italy). *Climate, 9*(7), 116. https://doi.org/10.3390/cli9070116

Ghofrani, Z., Sposito, V., & Faggian, R. (2017). A comprehensive review of blue-green infrastructure concepts. *International Journal of Environment and Sustainability, 6*(1).

Gunawardena, K. R., Wells, M. J., & Kershaw, T. (2017). Utilising green and bluespace to mitigate urban heat island intensity. *Science of the Total Environment, 584*, 1040–1055.

Herath, P., Thatcher, M., Jin, H., & Bai, X. (2021). Effectiveness of urban surface characteristics as mitigation strategies for the excessive summer heat in cities. *Sustainable Cities and Society, 72*, 103072. https://doi.org/10.1016/j.scs.2021.103072

Hu, L., & Li, Q. (2020). Greenspace, bluespace, and their interactive influence on urban thermal environments. *Environmental Research Letters, 15*(3), 034041.

Huang, H., Yang, H., Chen, Y., Chen, T., Bai, L., & Peng, Z. (2021). Urban green space optimization based on a climate health risk appraisal—A case study of Beijing city, China. *Urban Forestry & Urban Greening, 62*, 127154. https://doi.org/10.1016/j.ufug.2021.127154

Imran, H. M., Kala, J., Ng, A. W. M., & Muthukumaran, S. (2019). Effectiveness of vegetated patches as green infrastructure in mitigating urban heat island effects during a heatwave event in the city of Melbourne. *Weather and Climate Extremes, 25*, 100217. https://doi.org/10.1016/j.wace.2019.100217

Irmak, M. A., Yilmaz, S., Mutlu, E., & Yilmaz, H. (2020). Analysis of different urban spaces on thermal comfort in cold regions: A case from Erzurum. *Theoretical and Applied Climatology, 141*(3–4), 1593–1609. https://doi.org/10.1007/s00704-020-03289-y

Khorrami, M., & Malekmohammadi, B. (2021). Effects of excessive water extraction on ecosystem service groundwater: Vulnerability assessments using biophysical approaches. *Science of the Total Environment, 799*, 149304.

Lan, H., Lau, K. K., Shi, Y., & Ren, C. (2021). Improved urban heat island mitigation using bioclimatic redevelopment along an urban waterfront at Victoria Dockside, Hong Kong. *Sustainable Cities and Society, 74*, 103172. https://doi.org/10.1016/j.scs.2021.103172

Lehnert, M., Brabec, M., Jurek, M., Tokar, V., & Geletič, J. (2021). The role of blue and green infrastructure in thermal sensation in public urban areas: A case study of summer days in four Czech cities. *Sustainable Cities and Society, 66*, 102683. https://doi.org/10.1016/j.scs.2020.102683

Li, H., Meng, H., He, R., Lei, Y., Guo, Y., Ernest, A., et al. (2020). Analysis of cooling and humidification effects of different coverage types in small green spaces (SGS) in the context of urban homogenization: A case of HAU campus green spaces in summer in Zhengzhou, China. *Atmosphere, 11*(8), 862. https://doi.org/10.3390/ATMOS11080862

Lin, Y., Wang, Z., Jim, C. Y., Li, J., Deng, J., & Liu, J. (2020). Water as an urban heat sink: Blue infrastructure alleviates urban heat island effect in mega-city agglomeration. *Journal of Cleaner Production, 262*, 121411. https://doi.org/10.1016/j.jclepro.2020.121411

Lourdes, K. T., Gibbins, C. N., Hamel, P., Sanusi, R., Azhar, B., & Lechner, A. M. (2021). A review of urban ecosystem services research in Southeast Asia. *Land, 10*(1), 40.

Meili, N., Manoli, G., Burlando, P., Bou-Zeid, E., Chow, W. T. L., Coutts, A. M., et al. (2020). An urban ecohydrological model to quantify the effect of vegetation on urban climate and hydrology (UT & C v1.0). *Geoscientific Model Development, 13*(1), 335–362. https://doi.org/10.5194/gmd-13-335-2020

Mohajerani, A., Bakaric, J., & Jeffrey-Bailey, T. (2017). The urban heat island effect, its causes, and mitigation, with reference to the thermal properties of asphalt concrete. *Journal of Environmental Management, 197*, 522–538.

Nguyen, T. T., Meurk, C., Benavidez, R., Jackson, B., & Pahlow, M. (2021). The effect of blue–green infrastructure on habitat connectivity and biodiversity: A case study in the Ōtākaro/Avon River catchment in Christchurch, New Zealand. *Sustainability, 13*(12), 6732.

Perera, T. A. N. T., Nayanajith, T. M. D., Jayasinghe, G. Y., & Premasiri, H. D. S. (2021). Identification of thermal hotspots through heat index determination and urban heat island mitigation using ENVImet numerical micro climate model. *Modeling Earth Systems and Environment, 8*(1), 209–226. https://doi.org/10.1007/s40808-021-01091-x

Santamouris, M., & Osmond, P. (2020). Increasing green infrastructure in cities: Impact on ambient temperature, air quality and heat-related mortality and morbidity. *Buildings, 10*(12), 1–34. https://doi.org/10.3390/buildings10120233

Sanusi, R., & Bidin, S. (2020). Re-naturing cities: Impact of microclimate, human thermal comfort and recreational participation. In *Climate change, hazards and adaptation options* (pp. 545–562). Springer.

Sanusi, R., & Livesley, S. J. (2020). London plane trees (*Platanus* × *acerifolia*) before, during and after a heatwave: Losing leaves means less cooling benefit. *Urban Forestry & Urban Greening, 54*, 126746.

Sanusi, R., Johnstone, D., May, P., & Livesley, S. J. (2016). Street orientation and side of the street greatly influence the microclimatic benefits street trees can provide in summer. *Journal of Environmental Quality, 45*(1), 167–174.

Sanusi, R., Johnstone, D., May, P., & Livesley, S. J. (2017). Microclimate benefits that different street tree species provide to sidewalk pedestrians relate to differences in plant area index. *Landscape and Urban Planning, 157*, 502–511.

Shi, D., Song, J., Huang, J., Zhuang, C., Guo, R., & Gao, Y. (2020). Synergistic cooling effects (SCEs) of urban green–blue spaces on local thermal environment: A case study in Chongqing, China. *Sustainable Cities and Society, 55*, 102065. https://doi.org/10.1016/j.scs.2020.102065

Stanley, C. H., Helletsgruber, C., & Hof, A. (2019). Mutual influences of urban microclimate and urban trees: An investigation of phenology and cooling capacity. *Forests, 10*(7), 533. https://doi.org/10.3390/f10070533

Syafii, N. I. (2021). Promoting urban water bodies as a potential strategy to improve urban thermal environment. *Geographica Pannonica, 25*(2), 113–120. https://doi.org/10.5937/gp25-30431

Teoh, M., Shinozaki, M., Saito, K., & Said, I. (2021). Developing climate-led landscapes and greenery in urban design: A case study at Ipoh, Malaysia. *Journal of Asian Architecture and Building Engineering, 21*(4), 1640–1656. https://doi.org/10.1080/13467581.2021.1942881

United Nations. (2018). *Department of Economic and Social Affairs, Population Division 2019 World Urbanization Prospects: The 2018 Revision (ST/ESA/SER.A/420)*. United Nations.

Vo, T. T., & Hu, L. (2021). Diurnal evolution of urban tree temperature at a city scale. *Scientific Reports, 11*(1), 10491. https://doi.org/10.1038/s41598-021-89972-0

Wilis, K. J., & Petrokofsky, G. (2017). The natural capital of city trees. *Science, 356*(6336), 374–376. https://doi.org/10.1126/science.aam9724

Wu, C., Li, J., Wang, C., Song, C., Chen, Y., Finka, M., & La Rosa, D. (2019). Understanding the relationship between urban blue infrastructure and land surface temperature. *Science of the Total Environment, 694*, 133742. https://doi.org/10.1016/j.scitotenv.2019.133742

Xie, Q., & Li, J. (2021). Detecting the cool island effect of urban parks in Wuhan: A city on rivers. *International Journal of Environmental Research and Public Health, 18*(1), 1–15. https://doi.org/10.3390/ijerph18010132

Yilmaz, S., Külekçi, E. A., Mutlu, B. E., & Sezen, I. (2021). Analysis of winter thermal comfort conditions: Street scenarios using ENVI-met model. *Environmental Science and Pollution Research, 28*(45), 63837–63859. https://doi.org/10.1007/s11356-020-12009-y

Zhao, L., Lee, X., Smith, R. B., & Oleson, K. (2014). Strong contributions of local background climate to urban heat islands. *Nature, 511*(7508), 216–219.

Zhu, W., Sun, J., Yang, C., Liu, M., Xu, X., & Ji, C. (2021). How to measure the urban park cooling island? A perspective of absolute and relative indicators using remote sensing and buffer analysis. *Remote Sensing, 13*(16), 3154. https://doi.org/10.3390/rs13163154

Assessing the Community Participation in Ecotourism at Ulu Muda Forest Reserve, Malaysia

9

Zaiton Samdin, Siti Intan Nurdiana Wong Abdullah, and Thanam Subramaniam

Abstract

The government agencies have created opportunities in the provision of ecotourism to encourage the participation of local communities. The growing demand for ecotourism provides employment opportunities to raise their standard of living. The Ulu Muda Forest Reserve (UMFR) is one of the most attractive natural areas in Kedah, Malaysia, and has the potential for ecotourism. The present study analyses the involvement of the local community at UMFR based on their motivation, perceived benefits and conflicts. Using purposive sampling, an interview-based survey was performed among the local community involved in ecotourism. This study identified various socio-economic advantages stemming from ecotourism activities, but the lack of equal opportunities negated the influence on their participation. Their participation in ecotourism was highly influenced by the absence of conflict and driven by the presence of intrinsic motivation. Community participation does stimulate a sense of self-belongingness, thus increasing awareness towards conservation to sustain their source of living. Hence, community participation is one of the aspects that will

Z. Samdin (✉)
Institute of Tropical Forestry and Forest Products (INTROP), Universiti Putra Malaysia (UPM), Serdang, Selangor, Malaysia

School of Business and Economics, Universiti Putra Malaysia (UPM), Serdang, Malaysia
e-mail: zaisa@upm.edu.my

S. I. N. W. Abdullah
Faculty of Business and Communications, INTI International University, Nilai, Malaysia

T. Subramaniam
Centre for Research and Innovation in Tourism (CRiT) & School of Hospitality, Tourism and Events, Taylor's University, Subang Jaya, Malaysia

Z. Samdin et al. (eds.), *Tropical Forest Ecosystem Services in Improving Livelihoods For Local Communities*, https://doi.org/10.1007/978-981-19-3342-4_9

ensure the success of the ecotourism industry towards the sustainability of natural resources by providing relevant assistance in policy-making.

Keywords

Ecotourism · Participation · Involvement · Motivation · Ulu Muda Forest Reserve · Local community

9.1 Introduction

Ecotourism is considered a form of nature-based tourism that uses natural resources as a key component. Ecotourism contributes to the economic, social and environmental development of the local area. The ecotourism travel experience is particularly related to undisturbed and unpolluted natural areas such as national parks, protected areas, coastal and marine areas, wildlife sanctuaries and other protected flora, fauna and habitat areas (Cheung & Fok, 2014). Many nature-based activities related to ecotourism experiences—wildlife viewing, bird watching, hiking, trekking, nature education and more. These activities are considered attractive ecotourism products because they provide a unique experience for tourists. Therefore, ecotourism is considered an important source in generating economic benefits to the local community through tourism activities and spending.

The Ulu Muda Forest Reserve (UMFR) shows the richness of its genetic resources. The presence of a group of elephants bathing in the river and eating grass in the wild further adds to the uniqueness of this forest. The forest has also recorded hundreds of species of flora and fauna. It is also a habitat for endangered species such as tigers, tapirs and hundreds of bird species. Fish resources are also one of the attractions of anglers in addition to Tualang honey, which is an icon to the UMFR. Among the ecotourism attractions in the UMFR are the Hot Springs Saltlick, caves in Labua Cave and wildlife boat cruise activities. Tourists to UMFR can stay at the Earth Lodge, Kuala Labua. The journey to UMFR starts at Muda Lake Jetty. Then, tourists will take a boat for an hour and a half across lakes and rivers. Saltlick is also one of the factors leading to the attraction of large and small mammal species, with the largest number of saltlicks recorded in this forest. This diversity of genetic resources further contributes to a balance of biological control and, in turn, becomes one of the sources of ecotourism attractions in this forest.

The UMFR is located at the midpoint latitude of 6° 2′57.58″ N and longitude 100° 58′54.99″ E in Sik District, Kedah. It is located at an altitude of up to 1000 m in forests classified in the formation of lowland forests (Swaine, 1989). The UMFR is located in the northern state of Kedah, and it is the largest forest reserve in the state, accounting for about half of the forest area in Kedah. It is accessible from the jetty at Muda Lake, and the nearest village is Gubir. The UMFR consists of three main lakes, namely, Ahning Lake, Muda Lake and Pedu Lake. The largest lake is Pedu Lake, which covers an area of 15,500 ha. Muda Lake is smaller at 3382 ha, but it has a larger catchment area, and in this lake, there is a 6.6 km long tunnel that channels water from Muda Lake to Pedu Lake and is located in the UMFR. The UMFR is important as a water catchment area as water from the UMFR is channelled to three main dams, namely, Pedu Dam, Muda Dam and Ahning Dam.

Therefore, this study aims to identify community participation in ecotourism in UMFR. More specifically, this research examines the local community's participation and their motivation to involve in ecotourism. This study also identifies the benefits and conflicts that ecotourism presents to the local community. This study is essential in understanding the participation and involvement of communities to produce guidance for better management of ecotourism development in UMFR. The findings of this study will provide a functional guideline to authorities and organizations such as the Kedah state government, the Ministry of Tourism and Culture Malaysia, site operators or service providers of UMFR, travel agencies and tour guides. The results will also enrich knowledge and information on the UMFR, particularly through implementing policy related to the local community's participation and involvement in the management of ecotourism destinations.

9.2 Community Participation for Ecotourism Sustainable Development

Community participation is often viewed as a keystone in achieving the objectives of sustainable ecotourism development (Bello et al., 2016; Taylor, 1995). The involvement of the community in the ecotourism sector is believed to be voluntary, whereby local individuals participate in the decision-making and ecotourism activities to attain a sustainable livelihood (Kala & Bagri, 2018). In hindsight, community participation in ecotourism is capable of motivating locals to take on more responsibility in entrepreneurial ventures and collaborations with various internal and external stakeholders to promote a more holistic and sustainable tourism development (Idziak et al., 2015). Ultimately, this cohesive and integrated relationship increases the success in implementing sustainable ecotourism development (Dyer et al., 2007). Community participation is also viewed as essential in the collective response towards promoting social agendas and building resilience, especially during the pandemic (Marston et al., 2020).

Interestingly, past researchers have sought to understand community participation based on different levels or stages (Tosun, 2006; Kantsperger et al., 2019). Table 9.1 summarizes the various typologies of participation. One of the earliest and most widely used community participation models was developed by Arnstein (1969). According to Arnstein's (1969) model, community participation is induced by benefits and power of decision-making that involves eight levels. Hart (1992) developed his model based on Arnstein (1969) but in youth participation. In comparison, Wilcox's (1994) model was more consolidated with only five levels, namely, information, consultation, deciding together, acting together and supporting initiatives. In the studies of tourism, Jamal and Getz (1995) divided local community participation simply into two main categories—passive and active. They conceded that all forms of consultative and non-voluntary participation could be merged as

Table 9.1 Typology of participation (Arnstein, 1969; Hart, 1992; Wilcox, 1994; Jamal & Getz, 1995; Tosun, 1999; Shier, 2001; Baksh et al., 2012; Kantsperger et al., 2019)

Arnstein (1969)	Hart (1992)	Wilcox (1994)	Jamal and Getz (1995)	Tosun (1999)	Shier (2001)	Baksh et al. (2012)	Kantsperger et al. (2019)
Citizen control	Shared decision-making	Supporting initiatives	Active	Spontaneous participation	Shared power	Evaluating shared benefits	Legitimated
Delegated power	Youth lead and initiate action	Acting together				Managing	Collaborative
Partnership	Adult-initiated	Deciding together			Involved	Decision making	
Placation	Consulted and informed		Passive	Induced participation	Consulted		Consultative
Consultation	Assigned but informed	Consultation			Supported	Planning	
Informing	Tokenism	Information					Informative
Therapy	Decoration			Coercive participation	Listened to		Non-participative
Manipulation	Manipulation						

Source: Author compilation

passive participation. Meanwhile, active participation is where the locals are acting voluntarily and empowered to be involved to make their own decisions and have full control of all the ecotourism activities that would impact their livelihood, family, culture and community (Jamal & Getz, 1995). Similarly, in the tourism context, Tosun (1999) conceded that community participation is not universal and differs based on destination that can be divided into coercive, induced and spontaneous participation. Shier developed his model in 2001, which was revisited empirically on Nicaraguan children's participation in 2009. According to Shier (2001), participation goes through five stages—from being listened to, supported, consulted and involved to finally sharing power and responsibility for decision-making.

Alternatively, community participation was investigated from a tourism development perspective by determining the level of involvement in different stages in the process, namely, planning, decision-making, managing and evaluating benefit-sharing (Baksh et al., 2012). They confirmed that the majority of the villagers were involved, especially in the evaluation stage of the ecotourism activities in Tambaksari Village, Indonesia (Baksh et al., 2012). A similar empirical result applying Baksh's model was confirmed among the local community's participation in the ecotourism at the Pahang National Park, Malaysia (Tan et al., 2020) and at the Mwaluganje Elephant Sanctuary, Kenya (Kihima & Musila, 2019). These studies established that local community participation was restricted to permitting benefit-sharing, but they lacked participation and decision-making power during the planning and implementing stages (Kihima & Musila, 2019). A more recent study undertaken by Kantsperger et al. (2019) explored local participation in tourism development at an alpine destination in Germany. They focused on identifying the patterns behind the participation instead of the reasons. Through qualitative content analysis conducted, they categorized community participation into three main categories—namely non-participation, unofficial participation and official participation, which bear similarities to Tosun's (1999) model of participation in which the highest level of participation indicated a form of shared responsibility and power of decision-making.

Nonetheless, understanding local community participation for sustainable ecotourism development may vary across different cultures, regions and countries (Eshun & Tichaawa, 2020). Some researchers have argued that there is a difference between developed and developing countries due to the influences of social and economic power within the communities (Sebele, 2010; Liu, 2006; Njoh, 2002). According to Liu (2006), there was a lack of local participation in tourism projects in rural areas in Malaysia due to financial barriers even though most of the ecotourism projects involved mainly prominent and wealthy investors, including foreign private sectors (Wondirad, 2017). A recent study proposes that spontaneous, active and direct participation in ecotourism should be grounded on encouraging benefit acquisition and benefit-sharing among the local community (Sithole et al., 2021).

The level of participation is highly dependent on the situation, objectives, benefits, motivation, barriers and accessibility (Sosa & Brenner, 2021; Zhang et al., 2020). One of the well-known attempts to construct an integrated model to explain the antecedents of participation is the motivation–opportunity–abilities

(MOA) model, which is applied in this study (Ölander & Thøgersen, 1995). This concept was applied by Jepson et al. (2014) to reveal the factors that enable or impede participation in local events and festivals. Generally, 'motivation' is internal and personal satisfaction derived from sensory or emotional feelings that improve the overall experience and fulfil one's physiological needs (Lee & Wu, 2021). 'Opportunities' are presented through beneficial consideration and accessibility to participate in the planning and execution of ecotourism. The 'ability' concept in the MOA model is a conflicting factor that acts as barriers or constraints to perform the behaviour. This is often developed from one's awareness, experience, knowledge and limitation of resources (Jepson et al., 2014). Either way, the lack of understanding of community participation theory, especially in the ecotourism development context, implies the need for further investigation. The following sections discuss the three concepts underpinning the MOA model as the key antecedents to predict community participation in ecotourism at UMFR.

9.2.1 Motivation for Community Participation

Individual motivation as a proxy that guides behaviour has always been a relatively interrelated concept (Chou & Chang, 2017). Motivation has been regarded as an internal and personal process that leads to individual direction, effort and performance of behaviour (Schunk & DiBenedetto, 2020). Individuals who set goals of what they want to achieve can strengthen motivation. Therefore, it is argued that highly motivated individuals with a positive mental state are more likely to successfully participate in any activities (Fei-Hu et al., 2020). Local community participation in ecotourism projects which stems from voluntary actions based on internal desires and motivations as postulated in the MOA model (Ölander & Thøgersen, 1995) rather than the external pressure is said to be the highest achievement in the level of participation according to many previous typologies by Arnstein (1969), Tosun (1999) and Shier (2001).

Moreover, the presence of respected local leaders in community programmes can encourage a sense of ownership and participation among the local individual (Shim & Lee, 2003). On the other hand, the participation in ecotourism activities among youths was found to be motivated by their feelings to possibly take on the role of future leaders in this sector (Abukhalifeh & Wondirad, 2019; Selby et al., 2020). Additionally, local youths' participation is motivated by the need to protect their surrounding environment and improve their overall quality of life (Rahman & Singh, 2019; Agyeman et al., 2019). The overall sense of enjoyment, satisfaction and well-being relatively affects an individual's willingness to participate actively (Gal, 2017).

9.2.2 Benefits of Ecotourism on Community Participation

Largely, the coverage of ecotourism development contributes to the public welfare of local communities, but the degree and opportunity to gain benefits differ for each individual (Ma & Wen, 2019). The proponents of ecotourism as a sustainable development among communities argued that local participation, starting from the planning stage until the evaluation of shared benefits state, is necessary to ensure that each resident in the destination area would have the opportunity to gain the benefits (Simmons, 1994). Ecotourism has been highly publicized as a form of tourism activity that is more responsible and generates benefits for locals (Regmi & Walter, 2017). However, it was suggested that only a small and restricted number of locals had acquired the benefits of ecotourism-related activities in the community (Campbell, 1999).

The opportunities to benefit from ecotourism activities as reflected in the MOA model include employment priorities, provision of income, improved infrastructure and social welfare, which are said to depend on the ability of the locals to control how the resources are used and allocated to avoid any exploitation (Mensah, 2017). Furthermore, the level of community participation in ecotourism is fostered by how the project affects them at a personal level, especially by the perceived economic benefits of the project (Jepson et al., 2014). Access to ecological resources and priority in monetary gains remains a key concern in determining locals' support towards ecotourism projects (Akyeampong, 2011). A past study also suggested that many locals engage primarily in various ecotourism activities as a livelihood sustenance strategy that provides an opportunity while recognising that the benefits are somewhat unstable due to its seasonality (Harilal & Tichaawa, 2018). Hence, supporting and diversifying pathways that enhance primary income play a major role in inspiring the local community to become key players in a more equitable ecotourism development (Phelan et al., 2020). With greater diversification of ecotourism activities in the destination, it provides locals with more opportunities to gain economic benefits from participation through direct and indirect employment (Sosa & Brenner, 2021).

Moreover, the perceived social benefits are often associated with access to training in enhancing ecotourism-related skills and promoting interaction among stakeholders (Poudel & Joshi, 2020). Higher levels of social benefits would strengthen the local community's feelings of engagement and involvement in ecotourism activities (Sosa & Brenner, 2021). Thus, it is implied that local communities who can obtain a cumulative benefit from ecotourism are more likely to participate actively, especially at the planning and decision-making stage (Li, 2006).

9.2.3 Causes of Conflict on Community Participation

Despite the motivation and opportunities presented by ecotourism, several challenges could hinder the ability of local communities to participate, resulting in unsustainable development of this sector (Ölander & Thøgersen, 1995; Jepson et al.,

2014). For instance, conflicts may occur between local communities and other stakeholders such as the indigenous group, government agencies, private companies and tourists. In pursuing conservation and livelihood goals, the lack of clear policies and guidelines often constitute conflict (Dimitriou, 2017). According to Tosun (2000), such conflict may arise in the form of barriers, namely, operational, structural and cultural barriers. The operational barriers often occur due to the lack of proper coordination and collaboration. Kia (2021) concurs in the sense that poor coordination and insufficient information sharing between the stakeholders often take place in developing countries due to the remoteness of the ecotourism destination. Moreover, the bureaucracy and power struggle in the administration of the area also pose challenges to smoothly implementing ecotourism. Some local communities may be unaware of the legal requirements that may weaken their desire for participation in ecotourism development (Bluwstein, 2017). Meilani et al. (2019) also argued that many locals felt stifled due to their lack of operational skills, lower educational background and insufficient training that hindered their participation, especially at the decision-making levels.

The structural barriers refer to the physical aspect of externalities of the destination, such as the lack of proper infrastructure, poor safety conditions and unmanaged resources alongside the absence of effective marketing campaigns (Neger, 2021). Moreover, due to the increase in tourists that demanded better facilities, locals had to pay an inflated rate to use these facilities in the park (Acquah et al., 2017). According to a study by Kunjuraman and Hussin (2017), which was conducted in Malaysia, well-maintained infrastructure and recreational facilities that are accessible and affordable to the local community would increase their likelihood to participate in ecotourism such as homestay programmes.

The cultural barriers also play a major role in determining the local community's willingness to participate in ecotourism due to the lack of awareness among tourists (Bhan & Singh, 2014). Although local communities recognize the economic benefits of ecotourism, some are reluctant to welcome tourists for fear that it will cause negative impacts such as social illnesses and diminishing local cultures that inhibit their commitments to pursue this path (Safitri & Putra, 2018). For instance, in a study conducted by Jahan and Akhter (2018) in Bangladesh, local elders observed that the youths in their village started to change their dressing style that imitates the tourists. The lack of clear communication and opportunity for interactive ecotourism activities between host and visitor may lead to a clash of cultural values which can be enhanced through local cultural shows (Sangpikul, 2017). The lack of self-esteem among local community members on the uniqueness of their heritage leads to frustration, and the need for recognition may cause them to be disinterested to promote the ecotourism initiative (Zacarias & Loyola, 2017). Higher causes of conflict seem to lead to lower participation levels, yet the various conflicts discussed above differ between destinations. Hence, these barriers must be investigated and addressed to ensure higher levels of community participation in ecotourism development.

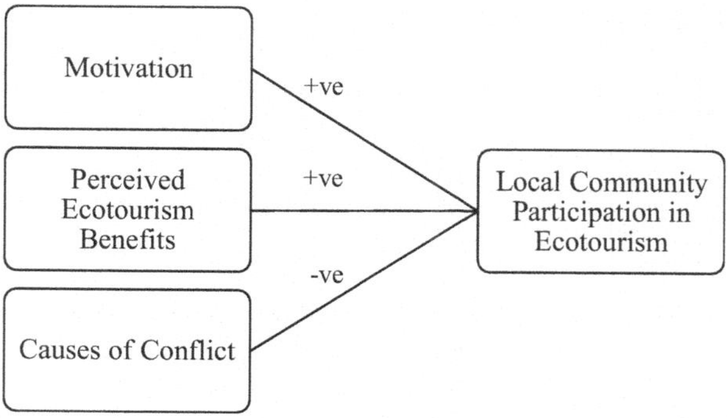

Fig. 9.1 Conceptual framework. (Source: Adapted from Jepson et al. (2014))

9.3 Conceptual Framework

Following the review of literature presented earlier, Fig. 9.1 shows the conceptual framework. Below conceptual framework reflects the main components of the MOA model to answer the research objectives. Although UMFR has a plethora of flora and fauna, the success of ecotourism depends on the participation and commitment of the local community. Strong participation from the local community would be a key factor in determining a well-managed ecotourism sector that will not jeopardize the natural resources of UMFR. Firstly, we hypothesized that higher levels of motivation will lead to higher levels of community participation in ecotourism activities. As discussed earlier, intrinsic motivation such as self-enjoyment, sense of achievement and overall well-being is a strong driver of their efforts and behaviour. Next, perceived ecotourism benefits present the 'opportunities' element of the MOA model, whereby the presence of various chances and the ability of the local community to grasp the opportunity to gain the benefits would have a positive and significant impact on their participation. However, the inability of locals to participate due to the existence of various conflicting reasons acts as a barrier and therefore is postulated to hinder their participation in the ecotourism activities. The items that are used to measure the respective variables are presented in Table 9.4.

9.4 Methodology

A quantitative research approach was adopted to examine the involvement of the local community in ecotourism and conservation activity in the UMFR. Saunders et al. (2012) indicated that the quantitative method should be associated with positivism, which focuses on examining existing theories while analysing the relationship between variables.

(a) Survey Instrument: Questionnaire

One set of questionnaire surveys applying the 5 Likert scale was used for data collection, whereby 1 is strongly disagree, 2 is agreed, 3 is neutral, 4 agrees and

Table 9.2 Reliability analysis

Variables	Cronbach's alpha	Number of items
Motivation	0.923	10
Benefits of ecotourism	0.940	14
Causes of conflict	0.966	16
Participation	0.901	7

5 strongly agrees. In this study, motivation, benefits of ecotourism and causes of conflict were used as independent variables, while local community participation in ecotourism was used as the dependent variable. Hence, sections A, B and C of the questionnaire contained statements related to the independent variables; Section D had statements for the dependent variable and Section E on demographic information. Throughout Section A–D, the variables are measured using a 5-point Likert scale. The statements were adapted from Kihima and Musila (2019), Noorhayati et al. (2015) and Adeleke and Nzama (2013).

(b) Research Sampling

The survey was conducted face-to-face between July 2021 and August 2021 at UMFR. A purposive sampling technique was executed in this study, mainly targeting only local communities currently actively involved in ecotourism at UMFR, whether directly or indirectly. The criteria of the respondents include (a) 18 years old and above and (b) actively involved in ecotourism in UMFR. Therefore, 32 qualified respondents were approached for this study. Quantitative data gathered via the questionnaire survey were analysed using Statistical Package for Social Sciences (SPSS).

9.4.1 Reliability Analysis

The correlation analysis is a statistical technique used to examine the strength and direction of the linear relationships between two variables (Pallant, 2013). Guildford's rule of thumb is used as a guideline to run the Pearson Correlation Coefficient Test to determine the strength of the association between the dependent variable and independent variables (Fadhil et al., 2007). Guildford's rule of thumb indicated that the R value less than 0.2 illustrates a negligible relationship, 0.20–0.40 as a low relationship, 0.40–0.70 as a moderate relationship, 0.70–0.90 as a high relationship and above 0.90 as a very high relationship. Cronbach's alpha coefficient was used, and the observed coefficient values for all variables in this study were above 0.90. The outcomes of the reliability test are presented in Table 9.2.

9.5 Results and Discussion

This section presents the findings of the study. It includes the description of the socio-demographic characteristics of the respondent; descriptive analysis of motivation, benefits of ecotourism and causes of conflict; and correlation analysis between motivation of ecotourism, causes of conflict and local community participation in ecotourism.

9.5.1 Socio-Demographic Characteristics of the Respondents

The survey was completed by 32 respondents ranging from 21 to 59 years old. These respondents' characteristics were divided into different categories as shown in Table 9.3. The respondents comprised 30.3% aged between 46 and 55 years old and 6.1% aged above 55 years old, 43.7% females and 56.3% males, 87.5% with secondary school education, 9.4% undergraduate and 3.1% postgraduate levels. Their monthly average income varied from less than MYR1500 to more than MYR5500. The largest proportion comprises 46.9% with middle income between MYR1500 and MYR2500, followed by 34.4% with income between MYR2500 and MYR3500. The highest income level proportion was 46.9%, whereas the lowest was 3.1%. The highest household number was between 4 and 6 (87.6%). In terms of occupation of the respondents, 75.0% were doing business or self-employed, and 15.6% were the private staff. Their role in ecotourism varies, including the provision of transportation (25.0%), followed by the lodging provider (21.9%), local F & B provider (18.8%) and small trading enterprise (15.6%).

9.5.2 Descriptive Analysis of Motivation, Benefits of Ecotourism and Causes of Conflict

(a) *Motivation*

Community motivation to participate in the ecotourism sector was determined using ten statements (Table 9.4).

In this study, 71.9% of the respondents agreed, and 28.1% strongly agreed that they enjoyed sharing the knowledge with others through participation in ecotourism activities. 68.8% agreed that participation has helped the locals to learn new things. The findings also indicated that 96.9% of the respondents agreed that they earn respect from others by contributing to the community through community participation. Meanwhile, 3.1% expressed a neutral stance to the statement. 96.9% agreed that they are motivated to participate in ecotourism activities because they felt it helped improve their quality of life. In comparison, 3.1% expressed a neutral stance to this statement. Hence, it can be concluded that the majority of the local communities were highly motivated to participate in ecotourism activities. They are mainly driven by their intrinsic needs, such as feelings of enjoyment, knowledge-sharing and personal

Table 9.3 Socio-demographic characteristics of the respondents

Demographic variable		Frequency	Percent
Age (years old)	18–25	7	21.2
	26–35	6	18.2
	36–45	8	24.2
	46–55	10	30.3
	>55	2	6.1
Gender	Male	18	56.3
	Female	14	43.7
Marital status	Married	24	75.0
	Single	8	25.0
	Others	–	–
Education level	Informal school	–	–
	Primary school	–	–
	Secondary school	28	87.5
	Certificate	–	–
	Undergraduate	3	9.4
	Postgraduate	1	3.1
Monthly income (MYR)	<MYR1500	4	12.5
	MYR1500–2500	15	46.9
	MYR2500–3500	11	34.4
	MYR3500–4500	–	–
	MYR4500–5500	1	3.1
	>MYR5500	1	3.1
Household number	<1	–	–
	1–3	3	9.3
	4–6	28	87.6
	7–9	1	3.1
	>9	–	–
Current work status	Government staff	2	6.3
	Private staff	5	15.6
	Business/self-employed	24	75.0
	Pensioner	1	3.1
	Unemployed	–	–
	Others: please specify	–	–
Role in ecotourism	Local tour guide	2	6.3
	Small trading enterprise	5	15.6
	Craftsman/handicraft maker	–	–
	Local tour operator/owner of travel agency	2	6.3
	Transportation provider	8	25.0
	Lodging provider	7	21.9
	Local F & B provider	6	18.8
	Local crop/livestock farmer	2	6.3

Table 9.4 Motivation influence the community participation

Statements	Strongly disagree (%)	Disagree (%)	Neutral (%)	Agree (%)	Strongly agree (%)
I find the community participation enjoyable	–	–	–	81.3	18.8
I enjoy sharing my knowledge with others via community participation	–	–	–	71.9	28.1
Community participation allows me to learn new things	–	–	–	68.8	31.2
Community participation enables me to become more proficient and enhance my expertise	–	–	–	65.6	34.4
I love to take part in discussions about community issues because I can help my community	–	–	9.4	68.8	21.9
I like to assist other community members with their questions/ inquiries	–	–	–	81.3	18.9
I earn respect from others by contributing to the community via participation	–	–	3.1	75.0	21.9
I actively participate in the activities organized by the community because it helps to improve our community's quality of life	–	–	3.1	84.4	12.5
I like to participate in community activities because it enhances my satisfaction	–	–	–	78.1	21.9
I enjoy participating in community activities because it allows me to do the activities on my own time	–	–	3.1	78.1	18.8

satisfaction. It means a person's enjoyment, intention to share knowledge with others and satisfaction encourage him or her to actively participate in ecotourism activities. Similarly, Liu et al. (2014), Moyle et al. (2010), Nault and Stapleton (2011) and Stylidis and Terzidou (2014) found economic benefits, experience stimulation, interest in environmental conservation and intention to improve the socio-economic motivating community involvement in the ecotourism activities. However, Hung et al. (2011) stressed the significant negative relationship between motivation and community participation.

(b) *Benefits of ecotourism*

In this study, the benefit of ecotourism was investigated through 14 statements, and the results are presented in Table 9.5.

The outcomes of the study indicated that 68.8% of the respondents agreed that the local community received priority in jobs and 31.2% of respondents strongly agreed with this. The findings revealed that 53.1% agreed that the

Table 9.5 Benefits of ecotourism to the community of UMFR

Statements	Strongly disagree (%)	Disagree (%)	Neutral (%)	Agree (%)	Strongly agree (%)
Community receive priority in jobs	–	–	–	68.8	31.3
Traditional skills of local people built	–	–	–	53.1	46.9
Local people have access to UMFR resources	–	3.1	15.6	62.5	18.8
Opportunity to sell local product available in UMFR	–	3.1	9.4	59.4	28.1
Income increases through assisting small local businesses	–	–	–	53.1	46.9
Provision of equipment for schools and clinics	–	–	–	59.4	40.6
Improvement of linking roads to communities	–	–	–	50.0	50.0
The local community receive access to the wildlife resources	–	–	12.5	62.5	25.0
Improvement in infrastructure (parking, toilet facilities, convenience shops) to communities	–	–	–	53.1	46.9
Income from practising local crafts, jungle trekking, wildlife watching, etc.	–	–	–	50.0	50.0
The community receives economic benefits from UMFR	–	–	12.5	56.3	31.3
Ecotourism courses and training provided to the local community	–	–	–	65.6	34.4
Opportunity to voice out local communities' opinions in planning and execution of ecotourism in UMFR	–	–	–	65.6	34.4
Training to enhance the local communities tour guiding skills and language	–	–	–	65.6	34.4

traditional skills of local people had improved while another 46.9% of respondents strongly agreed. This study revealed that 53.1% of them agreed that having small local businesses at UMFR helped enhance their income, and 46.9% strongly agreed. Meanwhile, 59.4% of the respondents agreed that the ecotourism activities helped them with equipment for schools and clinics, and another 40.6% strongly agreed with this. Also, all of the respondents agreed that ecotourism activities in UMFR had improved road access and the standard of living of the local community. 53.1% agreed, and 46.9% strongly agreed that ecotourism activities at UMFR helped to improve the infrastructure such as parking, toilet facilities and convenience shops. They also agreed with higher income obtained from various ecotourism activities such as selling local crafts, jungle trekking guiding, facilitating the wildlife watching activities and so on. In addition, 87.6% of the respondents agreed that the ecotourism activities did provide economic benefits. Overall, it can be concluded that ecotourism supports their socio-economic conditions by providing necessary facilities and infrastructures. Besides conservating the nature and environment, ecotourism stakeholders also need to prepare the necessary infrastructure and facilities such as toilets, accommodations, convenience shops, information counters, Wi-Fi services and so on which help to improve the socio-economics of local communities. Likewise, Rasoolimanesh et al. (2017) and Pasape et al. (2015) stressed that the community is interested to participate in ecotourism activities since it helps to generate the income benefits and social benefits and enhance the quality of life of residents by generating more job opportunities and supporting the infrastructure development. In addition, other benefits of ecotourism such as the refurbishment of local culture, building partnership, increasing economic benefits and improving economic and social conditions encourage community participation in the ecotourism activities (Stylidis & Terzidou, 2014; Moyle et al., 2010; Ali et al., 2022).

However, when inquiring about access to wildlife resources, only 87.5% of the respondents agreed with this statement. On the other hand, 62.5% agreed, and 25.0% strongly agreed that they have access to the wildlife resources at UMFR. Meanwhile, 12.5% selected a neutral option for this statement. Hence, it indicated that despite the various benefits obtained, not all local communities receive equal access to the wildlife resources at UMFR.

(c) *Causes of conflict*

The causes of conflict were examined through 16 statements, and the corresponding results are shown in Table 9.6.

In this study, 96.9% of the respondents disagreed that tourists do not respect the local culture during their visitation to UMFR, while 3.1% felt neutral for this statement. 96.9% disagreed that locals could not use the recreational facilities as opposed to 3.1% of the respondents who agreed with this statement. In addition, 93.8% disagreed that they were not allowed to voice their views in the decision-making process. Contrary, a minority of 3.1% agreed with this statement. On the ecotourism benefits, 96.9% disagreed that there is no benefit derived from ecotourism and conservation activities. Only 3.1% agreed with this statement. 96.9% of the respondents disagreed that ecotourism activities led to more crimes among local communities. Only 3.1% felt neutral on this statement. On limited employment opportunities, 96.9% disagreed. In the execution of park functions among locals, 96.9% of the respondents did not agree that there is any implementation of ecotourism park activities without consultation with the local communities. Around 93.8% of the respondents disagreed that ecotourism activities at UMFR affect the local communities' traditional livelihood. Only 6.2% agreed with this statement. Overall, it can be concluded that most of the local communities did not face high conflict since most of them felt that the ecotourism activities in UMFR were executed with the consultation of the local

Table 9.6 Causes of conflict in community participation

Statements	Strongly disagree (%)	Disagree (%)	Neutral (%)	Agree (%)	Strongly agree (%)
Tourists do not respect local culture	65.6	31.3	–	–	–
Locals are not allowed to access wildlife	34.4	40.6	21.9	3.1	–
Locals are not allowed to use recreational facilities	43.8	65.6	–	–	3.1
Opinion of locals not taken in decision-making	28.1	65.6	3.1	–	3.1
No benefits for locals from ecotourism and conservation	56.3	40.6	–	–	3.1
Increased crime in communities through ecotourism	53.1	43.8	3.1	–	–
Inadequate communication between park management and residents	31.1	62.5	3.1	3.1	–
Limited employment opportunities to the local communities	46.9	50.0	–	–	3.1
Lack of benefit-sharing opportunities to local communities	46.9	50.0	–	3.1	–
Limited opportunity for local communities to participate in the management operations	25.0	71.9	–	–	3.1
Restriction in accessing land, territory or road access of the park	21.9	65.6	9.4	3.1	–
Execute the park functionaries without consulting local communities	31.3	65.6	–	–	3.1
Hidden interests of the tourism authority on tourism development	25.0	68.8	3.1	3.1	–
Affecting traditional livelihood of the local communities	46.9	46.9	–	3.1	3.1
Resource use conflicts	12.5	65.6	18.8	3.1	–
Provision of compensation to local communities for losses	12.5	43.8	40.6	3.1	–

communities, thus not affecting their traditional ways of living. However, it is noteworthy that a small fraction did face conflict in sharing usage of recreational facilities and opportunities to participate in decision-making.

(d) *Community Participation in Ecotourism*
Finally, community participation in the ecotourism sector was tested through eight statements, and the descriptive results are presented in Table 9.7.

A total of 96.9% agreed that UMFR management had involved local communities in ecotourism planning. In this study, 92.8% of the respondents agreed that the locals

Table 9.7 Community participation in ecotourism

Statements	Strongly disagree (%)	Disagree (%)	Neutral (%)	Agree (%)	Strongly agree (%)
I am aware of the ecotourism activities and projects that take place in UMFR	–	–	–	78.1	21.9
UMFR management has involved local communities in its ecotourism planning	–	3.1	–	78.1	18.8
I feel personally involved in the decision-making process of ecotourism development in UMFR	–	–	6.3	75.0	18.8
I report any unsustainable practices within UMFR to the relevant agencies	–	–	3.1	68.8	28.1
I am currently participating in the ecotourism activities or projects in UMFR	–	3.1	3.1	68.8	25.0
I am actively involved in the implementation of various activities and projects in UMFR	–	3.1	9.4	68.8	18.8
I am an advocate in promoting ecotourism and conservation of UMFR	–	–	3.1	56.3	40.6

Table 9.8 Correlation result among motivation, benefits, conflicts and community participation

	Benefits of ecotourism	Cause of conflict	Motivation	Participation
Benefits of ecotourism	1.000			
Cause of conflict	**– 0.119**	1.000		
Motivation	**0.133**	**– 0.124**	1.000	
Participation	**0.068**	**– 0.716****	**0.401***	1.000

*Correlation is significant at the 0.05 level (1-tailed)
**Correlation is significant at the 0.01 level (1-tailed)

were involved in the decision-making processes of the ecotourism development in UMFR. Also, 96.9% agreed that they would take the responsibility to report any unsustainable practices within UMFR to the relevant agencies. The majority of the respondents (93.8%) agreed that, currently, they have actively participated in the ecotourism activities or projects organized by UMFR. 96.9% were also actively advocating the importance of ecotourism and conserving natural resources. The roles of ecotourism activities in boosting the tourists' experiences, promoting the local culture, improve the socio-economic of the local community motivating the community participation in ecotourism (Anup et al., 2015; Bhuiyan et al., 2011). A study by Salman et al. (2020) indicated that the motivation and encouragement of tourism stakeholders, especially from community leaders, led to high participation among residents in ecotourism activities.

9.5.3 Relationship Between Motivation, Benefits and Conflicts and Community Participation

The correlation analysis utilized in this study examines the relationship between motivation, benefits, conflicts and community participation in ecotourism activities. Table 9.8 shows the correlation coefficients of all the variables, and the outcome of the analysis indicated that only the cause of conflict and motivation were significant in influencing the local community's participation in ecotourism. The cause of conflict had the highest negative correlation coefficient (−0.716) towards community participation at a significance level of 0.00. Meanwhile, motivation showed a moderate correlation coefficient (0.401) with a significant level of 0.05. The results indicate that a lower conflict will enhance community participation. This supports the findings of Zacarias and Loyola (2017), who found that higher causes of conflict reduce the participation level among the community. Meanwhile, motivation had a positive and significant relationship with community participation. Hence, community participation will increase whenever there is a motivation to push or encourage the community. This was supported by the findings of Noorhayati et al. (2015), Mastura et al. (2020) and Sharpley (2014), which highlighted a significant relationship between motivation and community participation. However, the findings also illustrated a non-significant relationship between the benefits of ecotourism and community participation. The community does not seem to be influenced by the benefits that they will gain. Non-gaining or loss situation in the ecotourism sector does not seem to deter them from being involved in this sector. Contrary, it was against the findings of Wang et al. (2021) who indicated a positive and significant relationship between these variables.

9.6 Conclusion

This study aimed to understand the participation of the local community in ecotourism activities at UMFR by examining the influence of benefits of ecotourism, causes of conflicts and motivation. The absence of conflict would improve the local communities' participation. Hence, the local government plays a crucial role in ensuring that locals are given the opportunities and priority to participate at every level from planning to decision-making on the outcome of ecotourism benefits. This study found that the majority of the local communities were highly motivated to participate in ecotourism activities. They were driven by their intrinsic needs, such as feelings of enjoyment, knowledge-sharing and personal satisfaction. Hence, local government agencies in Kedah could motivate more locals to participate in ecotourism activities in UMFR by empowering them. For example, the Kinabatangan Tourism Cooperative (KOPEL) in Sabah, Malaysia, comprises 260 committee members from the local residents. These committee members are actively involved in the planning and implementation of various ecotourism activities. The initiative trained more than 300 local communities in operating homestays and also as tourist guides. As a key primary stakeholder, local communities should be given control over their livelihood and the resources at UMFR. Since the locals are highly motivated, they can act as role models in a mentor-mentee programme for other local community members who wish to participate in the ecotourism sector. Besides, the local communities indeed concur that ecotourism activities have brought them various benefits by improving their skills, increasing their income and providing numerous infrastructure improvements in UMFR.

However, findings indicated that not all local communities received equal access to the wildlife resources and opportunities to sell their local products at UMFR. A non-significant relationship was also found between the benefits of ecotourism and community participation. For these reasons, local communities are not opportunistic in that they look beyond the advantages. Hence, ecotourism can be promoted to be a part of their lifestyle. Due to this, local government agencies need to have clear guidelines that spell out the local communities' rights towards these resources in terms of access, usage and outputs, for instance, having a handbook of best practices management on responsible usage of natural resources such as land, water and energy, including protection of biodiversity species in the ecosystem. UMFR is not only a crucial water catchment area for the Northern states in Malaysia, but the local community depends on the forest for food security and ecotourism activities. Thus, the state and the federal government should revoke logging permits, enforce land-use control against illegal outsiders and collaborate with the local community on reforestation efforts in the gazetted forest reserve area. Community participation in ecotourism activities also led to the success and sustainability of activities and the conservation process. They also provide opportunities in the planning and decision-making process (Engku Nor et al., 2018).

9.7 Limitations and Future Studies

Due to the pandemic situation in the country, this study faced the limitation in obtaining higher numbers of respondents as the ecotourism sector had slowed down. Some of the local communities involved in ecotourism in UMFR had to temporarily find alternative sources of their income. Hence, to increase the response rate and enrich the findings, this study can be replicated during the post-pandemic to attract more local communities who are involved in ecotourism to participate. Lastly, the findings of this study provided insights on the participation and involvement of communities, thus guiding for better management of ecotourism development in UMFR. The results acted as a basic functional guideline to authorities and organizations such as the Kedah State Government, Ministry of Tourism and Culture Malaysia and site operators or service providers of UMFR to implement relevant policies to improve local community participation in ecotourism sectors.

References

Abukhalifeh, A. N., & Wondirad, A. (2019). Contributions of community-based tourism to the socio-economic well-being of local communities: The case of Pulau Redang Island, Malaysia. *International Journal of Tourism Sciences, 19*(2), 80–97. https://doi.org/10.1080/15980634.2019.1621521

Acquah, E., Nsor, C. A., Arthur, E. K., Boadi, S., et al. (2017). The socio-cultural impact of ecotourism on park-adjacent communities in Ghana. *African Journal of Hospitality, Tourism and Leisure, 6*(2), 1–14.

Adeleke, B. O., & Nzama, T. (2013). Assessment of community participation in ecotourism and conservation at Hhuhuwe-Umfolozi Park, South Africa. *Journal of Environment and Earth Science, 3*(3), 27.

Agyeman, Y. B., Yeboah, A. O., & Ashie, E. (2019). Protected areas and poverty reduction: The role of ecotourism livelihood in local communities in Ghana. *Community Development, 50*(1), 73–91.

Akyeampong, O. A. (2011). Pro-poor tourism: Residents' expectations, experiences and perceptions in the Kakum National Park Area of Ghana. *Journal of Sustainable Tourism, 19*(2), 197–213.

Ali, M. B., Quaddus, M., Rabbanee, F. K., & Shanka, T. (2022). Community participation and quality of life in nature-based tourism: Exploring the antecedents and moderators. *Journal of Hospitality & Tourism Research, 46*(3), 630–661.

Anup, K. C., Rijal, K., & Sapkota, R. P. (2015). Role of ecotourism in environmental conservation and socioeconomic development in Annapurna conservation area, Nepal. *International Journal of Sustainable Development & World Ecology, 22*(3), 251–258.

Arnstein, S. R. (1969). A ladder of citizen participation. *Journal of the American Institute of Planners, 35*(4), 216–224. https://doi.org/10.1080/01944366908977225

Baksh, R., Soemarno, H. L., & Nugroho, I. (2012). Community participation in the development of ecotourism: A case study in Tambaksari Village, East Java Indonesia. *Journal of Basic and Applied Scientific Research, 2*(12), 12432–12437.

Bello, F. G., Carr, N., & Lovelock, B. (2016). Community participation framework for protected area-based tourism planning. *Tourism Planning & Development*. Abingdon: Routledge Taylor & Francis Group, *13*(4), 469–485.

Bhan, S., & Singh, L. (2014). Homestay tourism in India: Opportunities and challenges. *African Journal of Hospitality, Tourism and Leisure, 3*(2), 1–8.

Bhuiyan, M. A. H., Siwar, C., Ismail, S. M., & Islam, R. (2011). The role of home stay for ecotourism development in East Coast Economic Region. *American Journal of Applied Sciences, 8*(6), 540–546.

Bluwstein, J. (2017). Creating ecotourism territories: Environmentalities in Tanzania's community-based conservation. *Geoforum, 83*, 101–113.

Campbell, L. M. (1999). Ecotourism in rural developing communities. *Annals of Tourism Research, 26*(3), 534–553.

Cheung, L. T., & Fok, L. (2014). The motivations and environmental attitudes of nature-based visitors to protected areas in Hong Kong. *International Journal of Sustainable Development & World Ecology, 21*(1), 28–38.

Chou, S. Y., & Chang, T. (2017). Being helped and being harmed: A theoretical study of employee self-concept and receipt of help. *Journal of Happiness Studies, 18*(6), 1573–1592.

Dimitriou, C. K. (2017). From theory to practice of ecotourism: Major obstacles that stand in the way and best practices that lead to success. *European Journal of Tourism, Hospitality and Recreation, 8*(1), 26–37.

Dyer, P., Gursoy, D., Sharma, B., & Carter, J. (2007). Structural modeling of resident perceptions of tourism and associated development on the Sunshine Coast, Australia. *Tourism Management, 28*(2), 409–422.

Engku Nor, K. E. H., Shuib, A., Kunasekaran, P., Johari, S., & Adam, S. (2018). Local community participation in ecotourism development in Tasik Kenyir, Malaysia. *World Applied Sciences Journal, 36*(1), 85–92.

Eshun, G., & Tichaawa, T. M. (2020). Community participation, risk management and ecotourism sustainability issues in Ghana. *Geo Journal of Tourism and Geosites, 28*(1), 313–331. https://doi.org/10.30892/gtg.28125-472

Fadhil, M. H., Azizan, N. A., & Shaharudin, R. S. (2007). The interaction between macroeconomic variables and the performance of mutual fund in Malaysia. In C. Gan (Ed.), *MFA 9th Conference*, 12 and 13 June 2007.

Fei-Hu, D., Yuan-Heng, G., & Qiao-Li, T. (2020). Effects of ecotourism travel motivation and social relations on mental state of the middle-aged and elderly group. *Revista de Cercetare si Interventie Sociala, 70*, 20–30. https://doi.org/10.33788/rcis.70.2

Gal, T. (2017). An ecological model of child and youth participation. *Children and Youth Services Review, 79*, 57–64. https://doi.org/10.1016/j.childyouth.2017.05.029

Harilal, V., & Tichaawa, T. (2018). Ecotourism and alternative livelihood strategies in Cameroon's protected areas. *EuroEconomica (Special Issue on Tourism in Africa), 1*(37), 133–148.

Hart, R. (1992). *Children's participation from tokenism to citizenship*. UNICEF Innocenti Research Centre.

Hung, K., Sirakaya-Turk, E., & Ingram, L. J. (2011). Testing the efficacy of an integrative model for community participation. *Journal of Travel Research, 50*(3), 276–288.

Idziak, W., Majewski, J., & Zmyslony, P. (2015). Community participation in sustainable rural tourism experience creation: A long term appraisal and lessons from a thematic villages project in Poland. *Journal of Sustainable Tourism, 23*, 1341–1362.

Jahan, K. M., & Akhter, H. (2018). Impact of ecotourism on the environment, society and culture of Ratargul Swamp Forest in Sylhet, Bangladesh. *Asian Journal of Environment & Ecology, 8*(1), 1–8.

Jamal, T. B., & Getz, D. (1995). Collaboration theory and community tourism planning. *Annals of Tourism Research, 22*(1), 186–204.

Jepson, A., Clarke, C., Ragsdell, G., et al. (2014). Investigating the application of the motivation–opportunity–ability model to reveal factors which facilitate or inhibit inclusive engagement within local community festivals. *Scandinavian Journal of Hospitality and Tourism, 14*(3), 331–348. https://doi.org/10.1080/15022250.2014.946230

Kala, D., & Bagri, S. C. (2018). Barriers to local community participation in tourism development: Evidence from mountainous State Uttarakhand, India. *Turizam: međunarodni znanstveno-stručni časopis, 66*(3), 318–333.

Kantsperger, M., Thees, H., & Eckert, C. (2019). Local participation in tourism development—Roles of non-tourism related residents of the Alpine Destination Bad Reichenhall. *Sustainability, 11*(24), 6947. https://doi.org/10.3390/su11246947

Kia, Z. (2021). Ecotourism in Indonesia: Local community involvement and the affecting factors. *Journal of Governance and Public Policy, 8*(2), 93–105.

Kihima, B., & Musila, P. (2019). Extent of local community participation in tourism development in conservation areas: A case study of Mwaluganje conservancy. *Parks, 25*(25.2), 47–56. https://doi.org/10.2305/IUCN.CH.2019.PARKS-25-2en

Kunjuraman, V., & Hussin, R. (2017). Challenges of community-based homestay programme in Sabah, Malaysia: Hopeful or hopeless? *Tourism Management Perspectives, 21*, 1–9.

Lee, C. J., & Wu, C. S. (2021). Recreation management research on recreational motivation and experience. *American International Journal of Business Management, 4*(6), 22–29.

Li, W. (2006). Community decision-making participation in development. *Tourism Management, 33*(1), 132–143.

Liu, A. (2006). Tourism in rural areas: Kedah, Malaysia. *Tourism Management, 27*, 878–889.

Liu, J., Qu, H., Huang, D., Chen, G., Yue, X., Zhao, X., & Liang, Z. (2014). The role of social capital in encouraging residents' pro-environmental behaviors in community-based ecotourism. *Tourism Management, 41*, 190–201.

Ma, B., & Wen, Y. (2019). Community participation and preferences regarding conservation and development policies in China's giant panda nature reserves. *Sustainability, 11*(18), 1–17.

Marston, C., Renedo, A., Miles, S., et al. (2020). Community participation is crucial in a pandemic. *The Lancet, 395*(10238), 1676–1678. https://doi.org/10.1016/S0140-6736(20)31054-0

Mastura, J., Shuhaida, M. N., Diana, M., Alireza, J., & Hashim, J. (2020). Motivational factors impacting rural community participation in community-based tourism enterprise in Lenggong Valley, Malaysia. *Asia Pacific Journal of Tourism Research, 25*(7), 799–812.

Meilani, M. M., Andayani, W., Faida, L. R. W., Maryudi, A., et al. (2019). Ecotourism in Sebangau National Park: An avenue to enhance local community livelihoods while protecting the ecosystem. *IOP Conference Series: Earth and Environmental Science, 399*(1), 012112. https://doi.org/10.1088/1755-1315/399/1/012112

Mensah, I. (2017). Benefits and challenges of community-based ecotourism in park-fringe communities: The case of Mesomagor of Kakum National Park, Ghana. *Tourism Review International, 21*(1), 81–98.

Moyle, B., Croy, G., & Weiler, B. (2010). Tourism interaction on islands: The community and visitor social exchange. *International Journal of Culture, Tourism and Hospitality Research, 4*(2), 96–107.

Nault, S., & Stapleton, P. (2011). The community participation process in ecotourism development: A case study of the community of Sogoog, Bayan-Ulgii, Mongolia. *Journal of Sustainable Tourism, 19*(6), 695–712.

Neger, C. (2021). Ecotourism in crisis: An analysis of the main obstacles for the sector's economic sustainability. *Journal of Ecotourism, 2021*, 1–23. https://doi.org/10.1080/14724049.2021.1942019

Njoh, A. J. (2002). Barriers to community participation in development planning: Lessons from the Mutengene (Cameroon) self-help water project. *Community Development Journal, 37*(3), 233–248.

Noorhayati, M. R., Hasmadi, I. M., Pakhriazad, H. Z., Khairil, A. W., et al. (2015). Identifying motivation factors of the participation of local community in tourism industry in National Park, Pahang, Malaysia. In *International Conference on Management and Technology in Knowledge, Service, Tourism & Hospitality*.

Ölander, F., & Thøgersen, J. (1995). Understanding of consumer behavior as a prerequisite for environmental protection. *Journal of Consumer Policy, 18*(4), 345–385. https://doi.org/10.1007/BF01024160

Pallant, J. (2013). *SPSS survival manual* (5th ed.). Open University Press.

Pasape, L., Anderson, W., & Lindi, G. (2015). Assessment of indicators of sustainable ecotourism in Tanzania. *Anatolia, 26*(1), 73–84.

Phelan, A., Ruhanen, L., & Mair, J. (2020). Ecosystem services approach for community-based ecotourism: Towards an equitable and sustainable blue economy. *Journal of Sustainable Tourism, 28*(10), 1665–1685. https://doi.org/10.1080/09669582.2020.1747475

Poudel, B., & Joshi, R. (2020). Ecotourism in Annapurna conservation area: Potential, opportunities and challenges. *Grassroots Journal of Natural Resources, 3*(4), 49–73. https://doi.org/10.33002/nr2581.6853.03044

Rahman, S., & Singh, H. R. (2019). The potential for marine ecotourism in Mukim Tanjung Kupang, Johor, Malaysia through capacity building of local communities via environmental education. *Samudera Journal of Maritime and Coastal Studies, 1*(1), 60–74.

Rasoolimanesh, S. M., Ringle, C. M., Jaafar, M., & Ramayah, T. (2017). Urban vs. rural destinations: Residents' perceptions, community participation and support for tourism development. *Tourism Management, 60*, 147–158.

Regmi, K. D., & Walter, P. (2017). Modernisation theory, ecotourism policy, and sustainable development for poor countries of the global South: Perspectives from Nepal. *International Journal of Sustainable Development and World Ecology, 24*(1), 1–14.

Safitri, D., & Putra, Z. F. F. (2018). Ecotourism of social culture aspect in Indonesia. *Proceeding International Conference on University and Intellectual Culture., 1*(1), 60–71.

Salman, A., Jaafar, M., Mohamad, D., & Malik, S. (2020). Ecotourism development in Penang Hill: A multi-stakeholder perspective towards achieving environmental sustainability. *Environmental Science and Pollution Research., 28*(31), 42945–42958.

Sangpikul, A. (2017). Ecotourism impacts on the economy, society and environment of Thailand. *Journal of Reviews on Global Economics, 6*, 02–312.

Saunders, M., Lewis, P., & Thornhill, A. (2012). *Research methods for business students*. Pearson Education Ltd..

Schunk, D. H., & DiBenedetto, M. K. (2020). Motivation and social cognitive theory. *Contemporary Educational Psychology, 60*, 101832. https://doi.org/10.1016/j.cedpsych.2019.101832

Sebele, L. S. (2010). Community-based tourism ventures, benefits and challenges: Khama Rhino Sanctuary Trust, Central District, Botswana. *Tourism Management, 31*, 136–146.

Selby, S. T., Cruz, A. R., Ardoin, N. M., et al. (2020). Community-as-pedagogy: Environmental leadership for youth in rural Costa Rica. *Environmental Education Research, 26*(11), 1594–1620. https://doi.org/10.1080/13504622.2020.1792415

Sharpley, R. (2014). Host perceptions of tourism: A review of the research. *Tourism Management, 42*, 37–49.

Shier, H. (2001). Pathways to participation: Openings, opportunities and obligations. *Children & Society, 15*(2), 107–117. https://doi.org/10.1002/chi.617

Shim, W. S., & Lee, Y. T. (2003). Residents' perceptions of government involvement impact and their attitudes towards government driven tourism development. *International Journal of Tourism Sciences, 3*(2), 133–150.

Simmons, D. G. (1994). Community participation in tourism planning. *Tourism Management, 15*, 98–108.

Sithole, N., Giampiccoli, A., & Jugmohan, S. (2021). Towards a spontaneous community participation model in community-based tourism. *African Journal of Hospitality, Tourism and Leisure, 10*(1), 222–237. https://doi.org/10.46222/ajhtl.19770720-97

Sosa, M. C., & Brenner, L. (2021). Factors of community participation that explain the benefits of ecotourism. *Revista de Turismo y Patrimonio Cultural, 19*(3), 453–476. https://doi.org/10.25145/j.pasos.2021.19.030

Stylidis, D., & Terzidou, M. (2014). Tourism and the economic crisis in Kavala, Greece. *Annals of Tourism Research, 44*, 210–226.

Swaine, M. D. (1989). Population dynamics of tree species in tropical forests. In *Tropical forest: Botanical dynamics, speciation, and diversity* (pp. 101–110).

Tan, H. T., Lim, E., & Mamat, M. P. (2020). Local community participation in ecotourism at National Park in Pahang, Malaysia. *The Malaysian Forester, 83*(2), 340–352.

Taylor, G. (1995). The community approach: Does it really work? *Tourism Management, 16*(7), 487–489.

Tosun, C. (1999). Towards a typology of community participation in the tourism development process. *Anatolia, 10*(2), 113–134. https://doi.org/10.1080/13032917.1999.9686975

Tosun, C. (2000). Limits to community participation in the tourism development process in developing countries. *Tourism Management, 21*(6), 613–633. https://doi.org/10.1016/S0261-5177(00)00009-1

Tosun, C. (2006). Expected nature of community participation in tourism development. *Tourism Management, 27*, 493–504.

Wang, M., Jiang, J., Xu, S., Guo, Y., et al. (2021). Community participation and residents' support for tourism development in ancient villages: The mediating role of perceptions of conflicts in the tourism community. *Sustainability, 13*(5), 2455. https://doi.org/10.3390/su13052455

Wilcox, D. (1994). *The guide to effective participation*. Partnership Books.

Wondirad, A. (2017). Who benefits from the ecotourism sector in Southern Ethiopia? *International Journal of Tourism Sciences, 17*(4), 276–297. https://doi.org/10.1080/15980634.2017.1384133

Zacarias, D., & Loyola, R. (2017). *How ecotourism affects human communities. In Ecotourism's promise and peril* (pp. 133–151). Springer. https://doi.org/10.1007/978-3-319-58331-0_9

Zhang, Y., Xiao, X., Cao, R., et al. (2020). How important is community participation to eco-environmental conservation in protected areas? From the perspective of predicting locals' pro-environmental behaviours. *Science of the Total Environment, 739*, 139889. https://doi.org/10.1016/j.scitotenv.2020.139889

Improving Livelihood of Sugar Palm Community in Malaysia

10

R. A. Ilyas, S. M. Sapuan, M. S. N. Atikah, M. R. M. Huzaifah, M. R. Ishak, and Z. Leman

Abstract

Engineering product development based on sugar palm tree in Kg. Kuala Jempol, Negeri Sembilan, Malaysia, is a joint venture project between Universiti Putra

R. A. Ilyas (✉)
School of Chemical and Energy Engineering, Faculty of Engineering, Universiti Teknologi Malaysia, Johor Bahru, Johor, Malaysia

Centre for Advanced Composite Materials (CACM), Universiti Teknologi Malaysia, Johor Bahru, Johor, Malaysia
e-mail: ahmadilyas@utm.my

S. M. Sapuan
Laboratory of Biocomposite Technology, Institute of Tropical Forestry and Forest Products (INTROP), Universiti Putra Malaysia (UPM), Serdang, Selangor, Malaysia

Advanced Engineering Materials and Composites Research Centre (AEMC), Department of Mechanical and Manufacturing Engineering, Universiti Putra Malaysia (UPM), Serdang, Selangor, Malaysia

M. S. N. Atikah
Department of Chemical and Environmental Engineering, Universiti Putra Malaysia (UPM), Serdang, Selangor, Malaysia

M. R. M. Huzaifah
Department of Crop Science, Faculty of Agricultural and Forestry Sciences, Universiti Putra Malaysia Bintulu Campus (UPMKB), Bintulu, Sarawak, Malaysia

M. R. Ishak
Department of Aerospace Engineering, Universiti Putra Malaysia (UPM), Serdang, Selangor, Malaysia

Z. Leman
Advanced Engineering Materials and Composites Research Centre (AEMC), Department of Mechanical and Manufacturing Engineering, Universiti Putra Malaysia (UPM), Serdang, Selangor, Malaysia

Z. Samdin et al. (eds.), *Tropical Forest Ecosystem Services in Improving Livelihoods For Local Communities*, https://doi.org/10.1007/978-981-19-3342-4_10

Malaysia (UPM) and Village Development and Security Committee (JKKK) of Kg. Kuala Jempol, Negeri Sembilan, with funding from the Ministry of Education Malaysia under the eighth National Blue Ocean Strategy (NBOS). This project is important in transferring university expertise to transform community. There were three main objectives, namely, to transfer the knowledge for the development of products based on sugar palm fibres, to transfer the knowledge for the development of products based on sugar palm tree and to help the local community in marketing the product. From this project, the village community is exposed to the potentials of sugar palm that is usually known for only sweetening beverages. At the end of this project, the community will continue the process of collecting and making products based on sugar palm trees and market them throughout Malaysia. The project has proudly developed 12 products based on sugar palm trees, namely, the sugar palm fibre, starch, roof, rope, brooms, brushes, bottlebrushes, vinegar, fruit, liquid sugar, fined sugar and block sugar. In a short time, all products can be realised along with packaging that can attract buyers. In addition, two products have received registered trademarks which are the sugar palm fibre and sugar palm starch.

Keywords

Sugar palm · Community · Sugar palm products · Fibre · Starch · Product development

10.1 Background

In 2015, the Ministry of Education Malaysia financed a community initiative at Universiti Putra Malaysia to produce products from sugar palm trees as part of the National Blue Ocean Strategy (NBOS) Initiative. According to Huzaifah et al. (2017a, 2017b), sugar palm tree has various names depending on place such as enau or kabung in Malaysia; gomuti, aren or kaong in Indonesia; palma azucarera in Spain; Lao in Sino-Tibetan; and many others. It belongs to the sub-family of Arecoideae and tribe of Caryoteae (Sanyang, 2015). Sugar palm tree has been chosen in this project because this tree is a multipurpose tree in which all parts of the tree can be used, including palm sap, trunk, fruits and leaves. The main purpose of NBOS initiative is to help the villagers so that the villagers can generate their own income.

In this project, the researcher was transferring information about the development of items made from sugar palm fibres, assisting the local community with product development and assisting the local community with product marketing. The project was carried out in a rural village in Bahau, Negeri Sembilan, Malaysia, called Kampung Kuala Jempol. The community was introduced to the potentials of the sugar palm, which was previously only known for producing palm sugar. Following the conclusion of the project, the community continued to collect and manufacture sugar palm tree-based items, which they sell across the country.

10.2 Products from Sugar Palm Trees

Coconut trees are widely recognised as multifunctional trees; however, few people are aware that sugar palm trees are also multipurpose. It's because almost every part of the tree (roots, leaves, stems, fibres, fruits, etc.) can be used for a variety of purposes and products (up to at least 60), either for traditional uses like sugar palm food products, *kolang kaling* (sugar palm fruits) (Fig. 10.1), vinegar or fibre products (such as fibre, roof, brush and broom) or for research purposes like starch and sugar palm fibre (Fig. 10.2), as well as base materials for a variety of structures. Researchers explored several uses of sugar palm products in the form of fibres (Fig. 10.3), foodstuff (Fig. 10.4) and other uses throughout the market investigation stage of the project (Fig. 10.5). Figure 10.5 shows tools constructed from sugar palm trunks and broomsticks made from sugar palm trees. Sugar palm seedlings in polybags are now available for purchase and planting (Fig. 10.5).

Fig. 10.1 Sugar palm fruits

Fig. 10.2 Sugar palm fibres and stem

10.3 Approaches

The NBOS project was completed in 1 year and 3 months, with a grant of USD 39,056.22 accepted (rate on 8 January 2018). The full sum was used to convey knowledge to the community and ensure its long-term viability. The funds were utilised to host community lectures and exhibitions and to design and create equipment such as a shed and a "small factory" to make products so that the local community can continue the initiative on its own in the future. The community had gotten a lot of orders, mostly from local markets, after running this initiative on their own for a year (project concluded in September 2016). Due to a lack of innovative techniques and procedures, particularly for mass manufacturing, the community has difficulty meeting demand. As a result, the same team of UPM researchers (who also worked on the first community project) launched a new community project called Knowledge Transfer Grant Scheme (KTGS) (with a grant approved of USD 6255.00; rate on 8 January 2018) to focus on project

Fig. 10.3 Different uses of sugar palm fibres

productivity enhancement to meet market demand and to assist villagers in marketing their products both locally and internationally.

10.3.1 Availability of Sugar Palm Trees

Sugar palm trees thrive in the undeveloped areas of the Jempol district's rural areas. In the village, some 1000 trees have been discovered in the wild. The sugar palm trees in Kampung Kuala Jempol were not widely and productively exploited by the inhabitants. Some of the locals did not use other portions of the sugar palm trees, who simply collected the sap and fruits for daily sustenance.

Fig. 10.4 Foodstuffs from sugar palm origin

10.3.2 Industry Visits

UPM researchers have put a lot of effort into the project, including making multiple trips to Indonesia, Sabah and Pahang, which have substantially aided the project's progress. The researchers went to Indonesia to learn more about the sugar palm industry from CV. Mulya Perkasa Company in Tasikmalaya, West Java Province, Indonesia, has extensive experience.

Expertise and information for the production of brushes, bottlebrushes, brooms, roofs and rope were brought back to Malaysia due to the tour (Fig. 10.6). Furthermore, information about sugar palm fibre as a roofing material was collected by visiting a community in Kampung Naga, Salawu District, Tasikmalaya Regency, West Java Province, Indonesia, that has utilised sugar palm fibre in roof making, which is very significant to emphasise in Malaysia. As a result of the development of holiday resorts that use traditional materials to build gazebos, cottages and chalets, demand for a traditional touch in building decorating is increasing in Malaysia. They became more motivated to manage the initiative after returning to Malaysia and sharing the information and products with the community. This is because the local

Fig. 10.5 Other uses of sugar palm tree

Fig. 10.6 Products from sugar palm fibres (rope, roof and broom)

community believed they could make it work based on the evidence presented and examples of firms that have been associated with the sugar palm. The problems in educating the community were obstacles that had to be overcome for the project to run smoothly.

Fig. 10.7 The manufacturing of sugar palm syrup (no. 1–4)

The second visit was to a company called Kebun Rimau Sdn. Bhd. in Tawau, Sabah, which produces sugar palm block and sugar palm syrup as its major products. Figure 10.7 depicts the stages of creating sugar palm syrup, from harvesting sugar palm sap to bottling sugar palm syrup. The equipment was created in partnership with Malaysia Agricultural Research and Development Institute and was designed by the firm (MARDI). After the sap is collected, it will be pre-heated to ensure that it is not readily damaged. The sap will then be filtered and sent into a second machine, which will treat it to make it denser and lengthen its shelf life. Following that, the syrup will be bottled and sold.

Benta, Kuala Lipis, Pahang, was the site of the third visit. The main product of this company is sugar palm block (Fig. 10.8). Almost most of the locals here make money by selling sugar palm blocks. Based on observation, the raw material for producing sugar palm blocks was gathered from wild sugar palm trees. They also make sugar palm blocks using a traditional manner.

Fig. 10.8 The sugar palm sap is cooked and poured into moulds

They climb the sugar palm tree early in the morning to get the sap. To remove any contaminants, the sap will be filtered before being placed in a large pan. The sap will next be boiled until it thickens and becomes viscous and then poured into the mould and left to harden before it cools. The final process is to pack the sugar palm blocks so they may be sold.

10.3.3 Technology Transfer (from Waste to Wealth)

Traditionally, agricultural waste has been used to dispose of the useless parts of sugar palm trees used for sap production. According to the use of sugar palm trees, the fibre and the tree trunk, leaves and fruit will be burned or left to decay, as illustrated in Fig. 10.9. UPM researchers devised a scheme to address this issue that converts these wastes into new goods, improving the locals' life in Kampung Kuala Jempol. The community was drawn to the project because of this concept. This project also included constructing a traditional building that houses the production process and distributes sugar palm goods. It is also hoped to increase the number of visitors to Kampung Kuala Jempol (Fig. 10.10).

UPM researchers briefed on the potential products from the sugar palm tree as part of this technology transfer. Aside from that, UPM taught the community how to make sugar palm-based products, including sugar palm block, sugar palm fruit, broom, bottlebrush, etc. In addition, UPM assisted in providing tools and machines for the manufacturing process.

The chief of Kampung Kuala Jempol praised the project and encouraged the community to take advantage of it. He also wants to promote this community as a

Fig. 10.9 The tree was burned to remove the fibre (*ijuk*); besides, the fruits and trunks were left to decay

tourist attraction in the Jempol district to fulfil the government's "One District, One Industry" initiative under rural industry.

Rural industry is also known as a traditional industry. This industry is a platform to produce any products or handicrafts produced by the village community who still use the traditional local methods inherited from their ancestors. Efforts to develop

Fig. 10.10 A series of demonstrations to the community on obtaining sugar palm starch and sugar palm fibres

local industries always get attention from the government because this industry is said to have a lot of importance. The government always gives support by giving incentives and capital to rural industrial entrepreneurs to develop in parallel with other sectors. Although the rural industry is considered small, it also contributes to its economy.

On the other hand, the existence of rural industry also provides employment opportunities to the children of the village itself. In turn, it can reduce the migration rate of rural people to the city centre. This situation can certainly reduce the city's population growth, and thus social problems can be avoided. In addition, the existence of rural industries can encourage residents to get involved in the business. Other than that, the rural industry run by the villagers can maintain the traditional artistic and cultural values inherited from time immemorial.

10.4 Products

Sugar palm fibre, starch, roof, rope, brooms, brushes, bottlebrushes, vinegar, fruit, liquid sugar, fined sugar and block sugar are the 12 goods based on sugar palm that has been successfully created (Fig. 10.11). All products with attractive packaging can attract buyers in a short period. In addition, two products, sugar palm fibre (Fig. 10.12) and sugar palm starch (Fig. 10.13), have registered trademarks (Fig. 10.13). Fibres (all fibre-based products), trunk (starches), flowers (sap for generating sugar) and fruits are the four primary sections of the sugar palm tree's products.

10.4.1 Sugar Palm Starch

Figure 10.13 depicts the steps involved in producing starch, also known as sago. Sugar palm sago starch manufacturing begins with the sugar palm tree being cut

Fig. 10.11 Twelve products from sugar palm tree

Fig. 10.12 Sugar palm fibre and sugar palm logo trademark™

Fig. 10.13 Sugar palm starch production and sugar palm logo trademark™

down. A chain saw is used to remove the inner part of the trunk. The sago material recovered during the extraction procedure is then soaked in water.

The sago detritus will be squeezed to extract the milk after a few hours and left overnight. The use of an extruder machine, as shown in Fig. 10.14, improves the extraction process of starch.

The milk will then separate into two layers: water and starch sediment. The starch will then be dried in direct sunshine after removing the water. The clump starch will be pulverised in the traditional way of making starch flour. Later, the original

Fig. 10.14 Sugar palm sago extruder machine for extracting starch

procedure is enhanced by using an extruder machine to extract the starch, similar to how coconut milk is made. The sugar palm starch can be purchased online using the Shopee application or directly at grocery stores. The price for 1-kg starch is between RM20 and RM30.

10.4.2 Products from Sugar Palm Sap

The sap is the major producer of the sugar palm tree. The production of sugar palm sap is not easy because it needs to be collected through the cut branches. The sellers or farmers who collect sugar palm sap must wait until the sap is full in the container; if the sap is left for too long it will turn into alcohol. A bottle of sugar palm sap was sold for RM4.

Other than that, sugar palm block, sugar palm syrup, fine sugar and vinegar (using the fermentation process), as well as bio-ethanol, can all be made from the sap. The price for sugar palm block varies between RM15 and 20/kg depending on the location. Sugar palm sap is obtained from male flower bunches because it produces a large amount of sugar palm sap of high grade. Sap can be yielded for up to 3 or 4 months. Sugar palm bunches that are shrink and dry indicate that sap production is coming to an end. Depending on the fertility of the trees, 4–5 L of sap can be gathered twice a day from each bunch. The number of male bunches present in

Fig. 10.15 Product derived from sugar palm sap

the tree can estimate the amount of sap present. If there are multiple male bunches on the tree, the sap from each one can be harvested simultaneously. Figure 10.15 depicts the products of sugar palm sap, whereas Fig. 10.16 presents the sugar palm block production flow.

10.4.3 Sugar Palm Fruits

Aside from the diversity of items made from sugar palm trees, they also produce fruits processed into food. According to some locations, the Malay community refers to this sugar palm fruit by its traditional name. However, some common names include "beluluk", "buah kabong" and "kolang-kaling", with "kolang-kaling" being the most popular in Indonesia. Because it is frequently employed in the culinary business, this fruit is one of the most popular side dishes in the Malay community (Fig. 10.17). The sugar palm fruit is oval, spiky and clear or white in colour. Sugar palm fruits can be purchased online or directly at a grocery with a price range of

Fig. 10.16 The production of sugar palm block starting from sap collection until packaging

Fig. 10.17 Sugar palm fruit

RM15–RM20/kg. The demand for sugar palm fruits will be increased when the Aidilfitri celebration comes.

10.4.4 Sugar Palm Fibres

Ijuk fibres are another local term for sugar palm fibre. It has a black tint and a diameter of 0.5 mm on average. This fibre can resist temperatures of up to 150 °C. These fibres' high strength and durability qualities make them multifunctional in harsh environments. Aside from that, it is also resistant to seawater. An ijuk fibre comprises hundreds of microfibrils held together by lignin, which gives the fibre its strength. On average, each sugar palm tree can generate about 15 kg of ijuk fibre. A machete is used to separate the ijuk from the tree. To make climbing a tall tree easier, a ladder is used, or notches are cut into the tree's height. The ijuk is then gathered before being packaged and ready for sale. Broom, brush, bottlebrush, roof and rope are examples of ijuk fibre products, as shown in Fig. 10.18. The product can be found at any grocery and the price is affordable (Table 10.1).

UPM is now undertaking research on sugar palm fibres (SPF) as a composite reinforcement material. Table 10.2 compares the tensile qualities of sugar palm fibres to those of other commercial natural fibres such as coir, cotton, kenaf and others.

Fig. 10.18 Products development from sugar palm fibres

Table 10.1 The product from ijuk fibre and their price according to Shopee application

No	Product	Price (RM)
1.	Broom	5.00–20.00
2.	Brush	1.20–3.60
3.	Bottlebrush	15.00–20.00
4.	Roof	12.00
5.	Rope (20 m)	5.00–10.00
6.	Ijuk fibre (1 kg)	20.00–30.00

Table 10.2 Tensile properties of sugar palm fibres and other commercial natural fibre

Fibre	Tensile strength (MPa)	Tensile modulus (GPa)	Elongation at break (%)	References
SPF/Kuala Jempol	233 ± 71.17	4.189 ± 1.61	20.6 ± 9.29	Huzaifah et al. (2017a, 2017b)
SPF/ Indonesia	219 ± 79.71	3.889 ± 1.78	20.4 ± 9.29	Huzaifah et al. (2017a, 2017b)
SPF/Tawau	211 ± 89.19	4.324 ± 1.15	15.8 ± 6.82	Huzaifah et al. (2017a, 2017b)
Cotton	287–597	5.5–12.6	3–10	Satyanarayana et al. (1990), Li et al. (2007)
Ramie	220–938	44–128	2–3	Li et al. (2007)
Hemp	550–900	70	1.6	Li et al. (2007)
Jute	393–800	10–30	1.5–1.8	Li et al. (2007), Rao et al. (2007)
Sisal	227–400	9–20	2–14	Rao et al. (2007), de Silva et al. (2008), Fávaro et al. (2010)
Kenaf	250	4.3	–	Lee et al. (2009)
Coir	108–215	4–6	15–40	Rao et al. (2007)

Misri et al. (2010) used a hand lay-up process to create a small watercraft utilising a hybrid of sugar palm fibre- and glass fibre-reinforced unsaturated polyester composites. Ijuk fibre is suited for water use due to its water resistance qualities, allowing it to survive in water for an extended period of time. However, because artificial fibre is stronger than a natural fibre, it is hybridised with glass fibre to improve mechanical durability. The experiments on sugar palm fibre undertaken by UPM researchers are listed in Table 10.3.

10.5 Equipment, Machines and Facilities Developed

A lot of equipment, machinery and facilities were produced as a result of this initiative, from harvesting to commercialising sugar palm goods. Figures 10.19, 10.20 and 10.21 depict some of the equipment and machines used in this project to convert sugar palm fibres into commercial products.

Table 10.3 List of selected studies conducted by UPM researchers on sugar palm fibre

Researchers	Fibres	Matrix	Process
Januar (2005)	Long sugar palm fibre	Epoxy	Hand lay-up
Suriani (2006)	Long sugar palm fibre	Epoxy	Hand lay-up
Dandi (2008)	Long sugar palm fibre	Epoxy	Hand lay-up
Leman (2009)	Long sugar palm fibre	Unsaturated polyester	Hand lay-up
Ishak (2009)	Long sugar palm fibre	Unsaturated polyester	Hand lay-up
Sahari (2011)	Sugar palm trunk fibre, sugar palm frond fibre, sugar palm bunch fibre, Ijuk fibre	Unsaturated polyester	Cold press
Sairizal (2011)	Long sugar palm fibre	Unsaturated polyester	Hand lay-up
Ishak et al. (2012)	Long sugar palm fibre	Unsaturated polyester	Cold press
Dandi (2012)	Short sugar palm fibre	High impact polystyrene	Hot press
Ibrahim (2013)	Long sugar palm fibre	Thermoset	Cold press
Sahari et al. (2013)	Sugar palm powder	Sugar palm starch	Hot press
Sanyang (2015)	Sugar palm cellulose	Sugar palm starch	Solution-casting technique
Ilyas et al. (2016, 2017a, 2017b, 2018)	Sugar palm nanocellulose	Sugar palm starch	Solution-casting technique

Fig. 10.19 Sugar palm fibre cleaned and isolated according to a certain grade

As illustrated in Figs. 10.22, 10.23, 10.24 and 10.25, a small factory was created to provide a better facility to accommodate all produced machines/equipment as well as working space for this project. This structure serves as a hub for the project and a collection, processing and commercialisation facility for sugar palm products. The

Fig. 10.20 Broom from sugar palm fibre

UPM team project developed the building, and the total cost was estimated to be around USD 20,000.00 using the grant funds. All project facilities, such as a loading bay, work area, office, room, bathroom and a separate small building for processing palm sugar and electric power and water supply, are available in the building.

Fig. 10.21 Bottlebrush and cleaning brush from sugar palm fibre processing

10.6 Conclusions

The goal of this community endeavour was to create sugar palm-based products. The National Blue Ocean Strategy (NBOS) initiative was supported by the Malaysian Ministry of Education (MOE). One of the goals of this project was to impart information about the production of products made from sugar palm fibres and get the community involved in the development and marketing of the product. This project aimed to introduce the community to the sugar palm tree's potential for goods other than sugar palm sap and fruit. This knowledge transfer programme included visits to three sugar palm-based businesses in Tasikmalaya, West Java Province, Indonesia; Tawau, Sabah; and Benta, Pahang, and training and equipment for the local population in Kampung Kuala Jempol, Negeri Sembilan. Sugar palm fibre, starch, roof, rope, brooms, brushes, bottlebrushes, vinegar, fruit, liquid sugar, finer sugar and block sugar are 12 goods developed due to this effort.

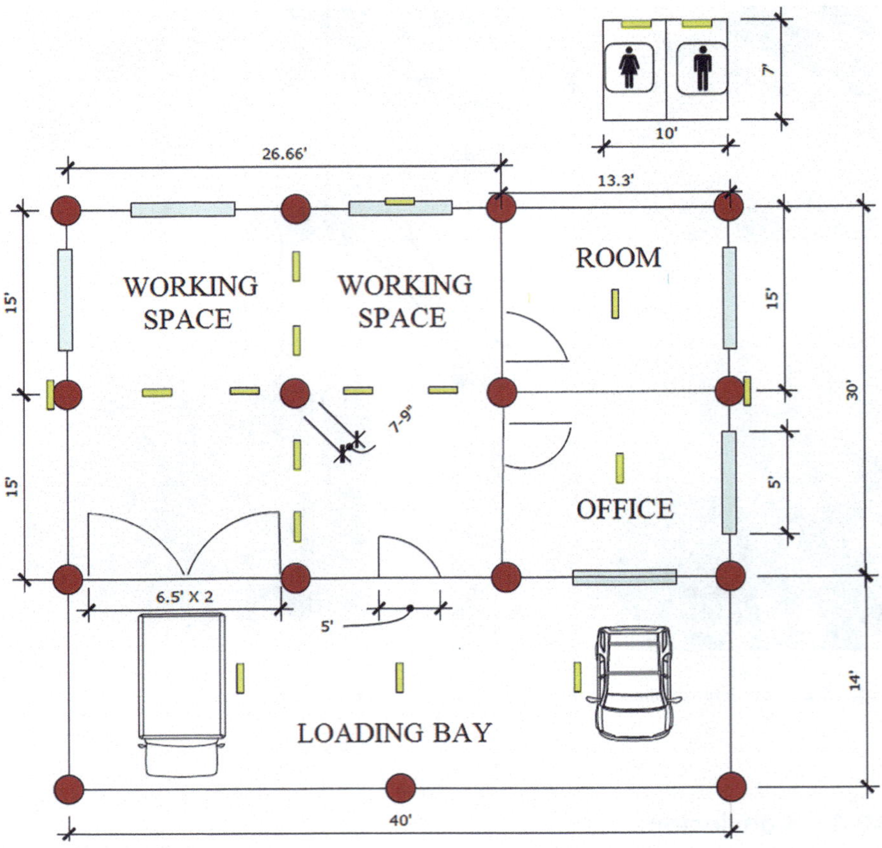

Fig. 10.22 Plan view of the building design

Fig. 10.23 Front view of the building design

Fig. 10.24 Isometric view of the building design

Fig. 10.25 The completed building photo

References

Dandi, B. (2008). *Mechanical properties of alkali-treated sugar palm (Arenga pinnata) fiber-reinforced epoxy composites* (Master of Science thesis, Universiti Putra Malaysia).

Dandi, B. (2012). *Mechanical and thermal properties of short sugar palm (Arenga pinnata Merr.) fiber-reinforced high impact polystyrene composites* (PhD thesis, Universiti Putra Malaysia).

de Silva, F. A., Chawla, N., & de Filho, R. D. T. (2008). Tensile behavior of high performance natural (sisal) fibers. *Composites Science and Technology, 68*(15–16), 3438–3443. https://doi.org/10.1016/j.compscitech.2008.10.001

Fávaro, S. L., Ganzerli, T. A., de Carvalho Neto, A. G. V., da Silva, O. R. R. F., & Radovanovic, E. (2010). Chemical, morphological and mechanical analysis of sisal fiber-reinforced recycled high-density polyethylene composites. *Express Polymer Letters, 4*(8), 465–473. https://doi.org/10.3144/expresspolymlett.2010.59

Huzaifah, M. R. M., Sapuan, S. M., Leman, Z., Ishak, M. R., & Maleque, M. A. (2017a). A review of sugar palm (*Arenga pinnata*): Application, fibre characterisation and composites. *Multidiscipline Modeling in Materials and Structures, 13*(4), 678–698.

Huzaifah, M. R. M., Sapuan, S. M., Leman, Z., & Ishak, M. R. (2017b). Comparative study on chemical composition, physical, tensile, and thermal properties of sugar palm fiber (*Arenga pinnata*) obtained from different geographical locations. *BioResources, 12*(4), 9366–9382.

Ibrahim, A. H. (2013). *Effects of flame-retardant agents on mechanical properties and flammability of impregnated sugar palm fiber-reinforced polymer composites* (Master of Science thesis, Universiti Putra Malaysia).

Ilyas, R. A., Sapuan, S. M., Sanyang, M. L., & Ishak, M. R. (2016). Nanocrystalline cellulose reinforced starch-based nanocomposite: A review. In *5th Postgraduate seminar on natural fiber composites* (pp. 82–87).

Ilyas, R. A., Salit, M. S., Sanyang, M. L., & Ishak, M. R. (2017a). Nanocrystalline cellulose as reinforcement for polymeric matrix nanocomposites and its potential applications: A review. *Current Analytical Chemistry, 13*, 203–225.

Ilyas, R. A., Sapuan, S. M., Ishak, M. R., & Zainudin, E. S. (2017b). Effect of delignification on the physical, thermal, chemical, and structural properties of sugar palm fiber. *BioResources, 12*(4), 8734–8754.

Ilyas, R. A., Sapuan, S. M., & Ishak, M. R. (2018). Isolation and characterisation of nanocrystalline cellulose from sugar palm fibers (*Arenga pinnata*). *Carbohydrate Polymers, 181*, 1038–1051.

Ishak, M. R. (2009). *Mechanical properties of treated and untreated woven sugar palm fiber-reinforced unsaturated polyester composites* (Master of Science thesis, Universiti Putra Malaysia).

Ishak, M. R., Sapuan, S. M., Leman, Z., Rahman, M. Z. A., & Anwar, U. M. K. (2012). Characterization of sugar palm (Arenga pinnata) fibres: Tensile and thermal properties. *Journal of Thermal Analysis and Calorimetry, 109*(2), 981–989.

Januar, P. S. (2005). *Tensile and flexural properties of Arenga pinnata filament (Ijuk filament) reinforced epoxy composites* (Master of Science thesis, Universiti Putra Malaysia).

Lee, B. H., Kim, H. J., & Yu, W. R. (2009). Fabrication of long and discontinuous natural fiber reinforced polypropylene biocomposites and their mechanical properties. *Fibers and Polymers, 10*(1), 83–90. https://doi.org/10.1007/s12221-009-0083-z

Leman, Z. (2009). *Mechanical properties of sugar palm fiber-reinforced epoxy composites* (PhD thesis, Universiti Putra Malaysia).

Li, X., Tabil, L. G., & Panigrahi, S. (2007). Chemical treatments of natural fiber for use in natural fiber-reinforced composites: A review. *Journal of Polymers and the Environment, 15*(1), 25–33. https://doi.org/10.1007/s10924-006-0042-3

Misri, S., Leman, Z., Sapuan, S. M., & Ishak, M. R. (2010, May). Mechanical properties and fabrication of small boat using woven glass/sugar palm fibres reinforced unsaturated polyester hybrid composite. In *IOP Conference series. Materials science and engineering* (Vol. 11, No. 1). IOP Publishing.

Rao, K. M. M., Prasad, A. V. R., Babu, M. N. V. R., Rao, K. M., & Gupta, A. V. S. S. K. S. (2007). Tensile properties of elephant grass fiber reinforced polyester composites. *Journal of Materials Science, 42*(9), 3266–3272. https://doi.org/10.1007/s10853-006-0657-8

Sahari, J. (2011). *Physio-chemical and mechanical properties of different morphological parts of sugar palm fiber reinforced polyester composites* (Master of Science thesis, Universiti Putra Malaysia).

Sahari, J., Sapuan, S. M., Zainudin, E. S., & Maleque, M. A. (2013). Mechanical and thermal properties of environmentally friendly composites derived from sugar palm tree. *Materials & Design, 49*, 285–289.

Sairizal, M. (2011). *Design and fabrication of a boat using sugar palm glass fiber reinforced unsaturated polyester composite* (Master of Science thesis, Universiti Putra Malaysia).
Sanyang, M. L. (2015). *Biodegradable composite films from modified sugar palm (Arenga pinnata (Wurmb) Merr.) starch for food packaging applications* (PhD thesis, Universiti Putra Malaysia).
Satyanarayana, K. G., Sukumaran, K., Mukherjee, P. S., Pavithran, C., & Pillai, S. G. K. (1990). Natural fiber-polymer composites. *Cement and Concrete Composites, 12*(2), 117–136. https://doi.org/10.1016/0958-9465(90)90049-4
Suriani, M. J. (2006). *A case study on tensile properties and morphology of Arenga pinnata fiber reinforced epoxy composites* (Master of Science, Universiti Putra Malaysia).

Local Community Involvement in Mangrove Forest Conservation and Edutourism in Kampung Sijangkang

11

Norfaryanti Kamaruddin, Mohd Hambali Mohd Jailani, and Zaiton Samdin

Abstract

Mangrove forests are biodiversity hotspots, and Malaysia has at least 70 mangrove species from 28 families, mainly from the Rhizophoraceae family. However, river pollution has posed a threat to the mangrove forest ecosystem and therefore should be conserved because it benefits various stakeholders. Edutourism is a good conservation model for mangrove forests given the strategic locations. The study interviewed the local community representatives to analyse the establishment of edutourism site called 'Taman Rekreasi Paya Bakau Kampung Sijangkang'. There are six sections in the interview: historical background, strategic partners, projects and activities, achievements, issues and challenges and way forward. We observed the progress of the activities carried out by the community in Kg. Sijangkang towards the development and its promotion as edutourism park. The programme is supported by the involvement of local authorities and other collaboration partners. In general, the socio-economic status of the community in Kg. Sijangkang has improved as a result of the economic contribution of educational and recreational activities in the Kuala Langat district. Our findings highlighted mangrove forest as an edutourism park in Kg. Sijangkang, which was developed by the local community. Our results showed the importance of self-driven community and teamwork to

N. Kamaruddin (✉) · M. H. Mohd Jailani
Institute of Tropical Forestry and Forest Products (INTROP), Universiti Putra Malaysia (UPM), Serdang, Selangor, Malaysia
e-mail: norfaryanti@upm.edu.my

Z. Samdin
Institute of Tropical Forestry and Forest Products (INTROP), Universiti Putra Malaysia (UPM), Serdang, Selangor, Malaysia

School of Business and Economics, Universiti Putra Malaysia (UPM), Serdang, Malaysia

Z. Samdin et al. (eds.), *Tropical Forest Ecosystem Services in Improving Livelihoods For Local Communities*, https://doi.org/10.1007/978-981-19-3342-4_11

progress and succeed in the conservation programme that could boost future socio-economic status. We also identified collaborative efforts as critical to the development of environmental conservation programmes.

Keywords

Mangrove forest conservation · Local community · Edutourism · Sustainable development goals · Socio-economic status

11.1 Introduction

Rural community development is one of the key agendas in the Sustainable Development Goals (SDG). The main strategies are to reduce poverty and hunger, promote gender empowerment and increase economic growth. Most rural communities are blessed with natural resources around them, which could be beneficial to their well-being and alleviate their socio-economic status. However, utilization of natural resources must be done responsibly.

Mangrove forests provide an important resource to society. They provide habitat for plants and fauna, protect the coast from erosion and provide a variety of items or materials for livelihood to the local community. Mangroves are forest formations with limited spatial distributions in the inter-tropical zone, between 30°N and 30°S latitudes, and roughly follow the 20 °C isotherms of seawater temperature (Giri et al., 2011; Spalding et al., 2010). Because of their high primary productivity and biomass, mangrove habitats have a high conservation priority (Komiyama et al., 2008; Ong et al., 2004) due to the unique presence of flora and fauna they shelter (Nagelkerken et al., 2008; Polidoro et al., 2010). Mangrove ecosystems have projected global biomass of 8.7 gigatons dry weight (i.e. 4.0 gigatons carbon) (Twilley et al., 1992), and these ecosystems perform vital natural functions such as fisheries production, coastal protection and habitat conservation (Duarte et al., 2013; Huxham et al., 2004).

To address the problem of mangrove forest degradation, significant parties, such as the village community, local government and other relevant stakeholders, must work together to restore, protect and manage mangrove forests. The local communities who are surrounded by the mangrove forests depend on the resources and its spin-offs economic activities. Therefore, the mangrove conservation programme has to be inclusive and involves a holistic approach to ensure that everyone benefits from the initiative.

Aside from the necessary involvement of the local community, existing tourist attractions must be given a sense of ownership so that the tourists' visit establishes positive feelings, especially if it contributes to their income. Community involvement in the development and maintenance of attractions will have a favourable impact on the long-term viability and the preservation of the tourist spot (Darius & Irena, 2018).

To achieve sustainable development goals, community development in edutourism is crucial. Tourism is a strategic instrument to realize Sustainable Development Goals (Ardika, 2018). It is to empower the local community through education and tourism to raise their income as well as their knowledge in environmental conservation activities. There is tremendous opportunity to be explored by the community when the area is surrounded by natural attractions, such as mangrove forests. The opportunity should be developed and sustained for the benefit of the society surrounding the area.

The effort to establish edutourism attraction needs sensible planning and consideration of numerous issues. Inhabitant readiness (Nabila & Yuniningsih, 2016), infrastructure readiness and management (Dewi, 2010) and edutourism tourist satisfaction (Darsiharjo & Saputra, 2016) are some of the major factors to be considered while designing edutourism in each place. The active role assumed by various stakeholders is critical to sustaining the edutourism attraction. The three important stakeholders are local authorities, learning institutions and the community (Arrasyid et al., 2021).

One of the approaches to utilise environmental services from coastal areas in a sustainable manner is mangrove forest ecotourism. Ecotourism development in mangrove habitats must be regulated to avoid hazards and negative environmental impacts, such as considering issues of suitability and environmental carrying capacity (Muhammad et al., 2010; Putro & Wasiq, 2015).

Fafurida et al. (2020) highlighted the strengths, weaknesses, opportunities and threats for the development of edutourism. The strengths include (1) conservation support, (2) local ecology support, (3) availability of flora and fauna vegetation, (4) abundant natural resources, (5) multiplier effect on the local community's economy, (6) natural laboratory for education and (7) aggressive promotion. The weaknesses are (1) lack of human capital in tourism management, (2) tourism education disturbing the environmental quality and (3) visitors' poor cleanliness practices. As for the opportunity, they are (1) alternative to nature tourism, (2) government's policy support, (3) opening investment opportunity, (4) promoting a back-to-nature lifestyle and (5) gaining society's support. The threats identified are (1) trade-offs between environmental damage and mass tourism promotion, (2) visitors who are not pro-environment, (3) imbalance development of tourism area and (4) illegal activities surrounding the attraction area.

The purpose of this chapter is to assess the development of edutourism attraction by the local community of Kg. Sijangkang and analyse their conservation model of mangrove forests. The assessments were carried out through interviews and surveys. The interview covered historical background, strategic partners, projects and activities, achievements, issues and challenges and way forward. The survey is to identify the perceptions towards the benefits of establishing the Taman Rekreasi Paya Bakau Kampung Sijangkang. This chapter also highlights the establishment, development process and their achievements.

11.2 Methodology

11.2.1 Study Area

Kampung Sijangkang is one of the villages located approximately 12 km from Klang to Banting in Kuala Langat district, Selangor, Malaysia. Sijangkang is the nearest town (Fig. 11.1). The name Sijangkang derives from a tree that resembles a rubber tree, specifically the Jangkang tree. It is a forest tree in which the seeds are propagated by expulsion and can be also found in some places in Indonesia. Historically, the Jangkang trees perish in a tributary river that was later named the Jangkang river, which constitutes along with other tributaries of the Langat River. The name is also culturally associated with Chinese since death is known as 'si', which eventually became known as 'SIJANGKANG'.

It is estimated that the village was first opened around 1907. Part of the villagers is originally from Indonesia, i.e. Java, while others came from other villages that have been open for a long time such as Kampung Beting Tengah and Kampung Pendamar. They landed from Sungai Jangkang from a small base found by the riverbank. In the 1930s, basic facilities, such as roads, began to be built connecting one place to another. The main crops in Kampung Sijangkang at that time were rubber and coffee.

Because the mangrove forest in Kg. Sijangkang offers a natural landscape that can attract the attention of the tourists in Selangor and Kuala Lumpur in particular, it has the potential as an ecotourism attraction. It is situated 55 km from the capital city of Kuala Lumpur. The local community has committed to management efforts,

Fig. 11.1 Map of Taman Rekreasi Paya Bakau Kampung Sijangkang generated from Google Map

resulting in the establishment of mangrove forest ecotourism in Kg. Sijangkang. It should be promoted as a leading tourist attraction for environmental conservation in the Klang Valley region.

11.2.2 Data Collection

We interviewed the head of the NGO managing the Taman Rekreasi Paya Bakau Kampung Sijangkang in Kuala Selangor. Three main subjects were discussed during the interview: the motivation to start the edutourism, the process of developing the edutourism and the achievements to date. The interview was done through a conference call due to the travel restrictions during the COVID-19 pandemic.

Eighty-five villagers were surveyed for their response on their observations of the benefits of conserving the mangrove forest. Five questions in the questionnaire collected respondents' perceptions towards the benefits of the edutourism program of Taman Rekreasi Paya Bakau Kampung Sijangkang. The respondents were the members of the association managing the Taman Rekreasi Paya Bakau Kampung Sijangkang.

We analysed the activities that have been carried out at the edutourism spot, Taman Rekreasi Paya Bakau Kampung Sijangkang from 2016 to 2021. The activities were categorised based on the main activities, sub-activities, total number of activities and partnerships.

11.3 Results and Discussion

This chapter is to assess the development of edutourism attraction by the local community of Kg. Sijangkang and analyse their edutourism programme of mangrove forests, how they started and their development process and success, and to assess the perceived impacts to local communities. Specifically, the focus was to highlight mangrove management models that have a high prospect of acceptance and adoption by local communities. This will ensure mutual future protection and community prosperity in various aspects.

We interviewed the head of the NGO managing the Taman Rekreasi Paya Bakau Kampung Sijangkang, Kuala Selangor, and surveyed 85 villagers to obtain their responses on the benefits of conserving the mangrove forest. We observed a successful and sustainable mangrove edutourism model in Kg. Sijangkang, based on the three domains: motivations, development process and achievements.

11.3.1 Motivations

Based on the interviews, the main motivation is to clean the riverside area that has become a trash dumping site. A low level of awareness among the villagers resulted in resistance among them at the beginning of the programme. It started in 2015, with communal activities among the villagers to clean the riverside occasionally until it becomes a routine. Then came the idea of developing and establishing the area into edutourism. The conservation and management of natural resources depend on a variety of local communities' socio-economic situations (Bourn & Brown, 2011; Crona & Hubacek, 2010; Weeks et al., 2010). Conservation is a way to get the community involved in mangrove recovery and management activities as a preservation effort (Wardhani et al., 2020). This effort can be supported by ecotourism activities which could improve the community welfare.

Mangrove degradation negatively affects the livelihoods of communities that depend heavily on mangrove ecosystems. The disruption consequently led to the emergence of a local champion to initiate mangrove rehabilitation efforts which initially received little attention from most communities. After some initial successes, the efforts of the local champion were subsequently replicated by other members of communities, triggering a bigger scale of the mangrove conservation area (Kusmana & Sukristijiono, 2016).

Successful management in mangrove forest conservation and restoration requires commitment from local communities, governments and stakeholders. This is because the farther humans are from nature, the less concerned they are about it (Wardhani et al., 2020). The community adopts an environmentally friendly development approach that had economic value and was able to improve the community welfare (Singgalen, 2020).

A pro-environmental attitude affects restoration activities (Byrka et al., 2010). Perception, attitude and behaviour of the surrounding community largely determine the condition of a forest area today and in the future (Renjaan et al., 2020).

Replanting activities is one of the factors that play a role in increasing mangrove land cover (Wardhani et al., 2020). Replanting involves managing the nursery, especially at the seedling stage, and maintaining planted trees until they reach a mature age. These processes could involve other stakeholders, learning institutions, private companies, NGOs and local authorities. The replanting programme is a way forward to continue the engagement between local communities and the stakeholders and sustain the mangrove land cover.

Mangrove rehabilitation in the community generally involves a bottom-up process led by the local indigenous group (Sadono et al., 2020). The efficiency of the bottom-up scheme compared to top-down is documented in several studies, particularly for natural resource management, where local indigenous communities reportedly demonstrate a strong commitment to sustaining the mangrove forests.

Several studies have reported the efficacy of bottom-up schemes over top-down schemes, particularly in natural resource management, where local indigenous communities demonstrate a strong commitment to conserving mangrove forests.

Sustaining the economic aspect is a great motivation for societies to initiate conservation activities. For instance, Aheto et al. (2016) highlighted the community's strong motivation to recover the mangroves that have been lost to degradation, which threatened their source of livelihood due to the sharp decline in fisheries productivity.

11.3.2 Development Process

The edco-tourism (education in ecotourism) concept is based on potential tourism resources. It is a concept that combines educational and environmental as the basis of local tourism management (Rakhmanissazly et al., 2018). Meanwhile, ecotourism directs tourism towards exotic, natural environments and supports conservation efforts. According to The International Ecotourism Society (TIES:2000), 'Ecotourism is not only a tourist but also a responsible natural tourism trip by conserving the environment and improving the welfare of local communities'.

Mangrove ecotourism has been identified as an ideal approach to provide mangrove education through ecotourism for people to understand mangroves' role in coastal protection. Mangrove conservation is an effort to restore the functions of mangroves as protectors and mitigate climate change. A study also showed the substantial contribution of mangrove ecotourism to raising public awareness, especially on the importance of mangroves in coastal areas, and the ecological system of the mangroves has been highlighted as a potential economic resource to the coastal community and the local government (Thompson & Friess, 2019). The implementation of ecotourism has a multiplier effect on educating and preserving mangrove vegetation.

The recommended strategies in ecotourism development include (1) complementing the infrastructure supporting ecotourism activities such as roads and bridges, (2) intensive promotion and (3) synergizing ecotourism with Kei traditional culture (Renjaan et al., 2020). Rehabilitation and management of mangrove forests need to be done to prevent the destruction of mangrove forests by involving various parties, including the village community, local government and other stakeholders.

Co-management is an approach to engaging the community in mangrove management that can mediate the management conflict between the local community and the government (Fatimatuzzahroh et al., 2020). The characteristics of community-based ecotourism based on Walter (2011) include (1) the principle of local community participation, monitoring and regulating or initiating ecotourism activities, (2) focusing on environmental conservation and the interests of local communities, (3) promoting customary law and local culture (local wisdom) to a certain extent and (4) promoting local human rights and indigenous peoples over traditional territories and resources (Reimer & Walter, 2013).

Numerous programs and activities have been carried out at the Taman Rekreasi Paya Bakau Kg. Sijangkang between 2016 and 2021. The details of the activities are shown in Table 11.1.

Mangrove recovery remains based on an environmentally friendly management approach through a series of innovative development (Gualard et al., 2018). Integrated efforts are needed in managing, conserving and protecting mangrove forests. Therefore, a mangrove management plan requires a greater effort to connect the stakeholders (Canty et al., 2018).

Singgalen (2020) emphasised the community's ability to maintain and utilize the mangrove forest using a community-based ecotourism approach to long-term

Table 11.1 Main activities carried out by the local community from 2016 to 2021

Main activity (2016–2021)	Sub-activities	No. of activities (as of July 2021)	Partnerships
Community engagement and replanting programme	Motivational school camps, company retreat programme, university's community engagement programme and international agencies' visits	150	Schools, private companies, public-listed companies, NGOs, universities and colleges and international agencies
Communal work	Festivities gatherings, cleaning and upgrading work surrounding the conservation area	21	Local authorities
TV/radio promotion	Programmes launching, marketing of the tourism spot and sharing sessions	8	Local TV and radio station

livelihood sustainability. The efforts to preserve the environment are supported by local government policies, institutions and the community's culture. Results indicate that the conservation of mangrove forests through community-based ecotourism enhances the sustainability of livelihood for the community. Therefore, the optimization of mangrove forest resources through a community-based ecotourism approach is relevant and practical.

Corporate Social Responsibility (CSR) programmes from the industrial sector cannot offset environmental losses. Environmental damages from the industrial sector have become a stimulus for the community to adopt environmentally friendly rural development approaches. The commitment of the company-community relationship is mutual and should be encouraged in creating positive outcomes and solutions for both parties. Value creating through CSR programmes may help nature preservation, but it is more helpful in mitigating negative perceptions from the community. It is about making the company has a humanism point of view by taking social responsibility seriously into the business model (Rakhmanissazly et al., 2018).

As with other reforestation efforts, mangrove reforestation is almost impossible to implement without the participation of communities since they have an essential role as the main actor to conduct reforestation activities, starting from planning, implementing to long-term monitoring (Budiharta et al., 2016). The researchers emphasized the importance of communities' participation as one of the factors that determine the success of reforestation activities in the mangroves (Eddiwan, 2018; Le et al., 2012; Wylie et al., 2016).

The awareness of communities regarding the benefits of mangrove forests can accelerate the execution of programmes. Moreover, they will also help in maintaining the sustainability of the mangroves after reforestation. It can be conducted in two stages: literature review and deep interview (Bengtsson, 2016).

Fig. 11.2 Champion in *Selangor Sungai Angkat* state-level programme

11.3.3 Achievements

A series of achievements have been made by the Kg. Sijangkang community throughout the year of 2019. The achievements proved that the community has successfully implemented their various conservation activities with the stakeholders. It also demonstrates their fruitful effort to ensure they are protecting their valuable gems. The awards and recognitions further encourage and excite them to preserve the mangrove forest as the edutourism spot.

In 2019, they were awarded as a champion in the Selangor state-level program of 'Sungai Angkat' (Fig. 11.2). The Sungai Angkat program is about river conservation, keeping it clean from any pollution. The purposes of the programme include nurturing and promoting public awareness and river conservation among the community; monitoring the quantity and quality of rivers from physical, chemical and biological pollutions; and encouraging community involvement in the districts (by NGOs, schools and others) by developing environmentally friendly infrastructure, go-green activities to improve water quality and ecotourism.[1]

They also have been listed in the Top 5 of a National Scroll of Honour for Human Settlements award during the National World Habitat Day and National Recycling Day in 2019 (Fig. 11.3). The National Scroll of Honour Award for Human Settlements was initiated to recognise the contribution made by society (public and private) to the sustainable development of human settlements in Malaysia. This award is fashioned on the UN-Habitat Scroll[2] of Honour Award, which acknowledges outstanding contributions to sustainable development, including improving the quality of urban life for all. It also celebrates Malaysia's success

[1] https://www.luas.gov.my/v3/my/luas/pengurusan/jawatankuasa/jawatankuasa-pelaksana-pemuliharaan-sungai-sungai-di-selangor

[2] https://www.un.org/en/observances/habitat-day

Fig. 11.3 National Scroll of Honour for Human Settlements 2019

stories of outstanding contributions in various domestic initiatives in numerous fields, such as shelter provisioning, innovation in solid waste management and sustainable urban planning, highlighting the plight of the homeless, improving the quality of urban life and providing innovative solutions on housing development. To be listed in the Top 5 for this honourable award is a significant achievement to the Taman Rekreasi Paya Bakau Kampung Sijangkang community. They expressed their gratitude as follows:

> "Thank you to MDKL for trusting us by nominating Sijangkang Mangrove Recreation Park. Hopefully, this award will raise the spirit of the volunteers to continue to serve the environment, especially the Mangrove Forest. Thank you to all the contributors and visitors because you helped us in keeping the Sijangkang Mangrove Recreation Park alive."—Sijangkang Mangrove Recreation Park Chairman

The Green Incentives Award was awarded by the Federal Department of Town and Country Planning (PLANMalaysia) to Kg. Sijangkang community during the World Town Planning Day Selangor state level in 2019 (Fig. 11.4). The award is part of their initiative to green the city and intensify the effort in supporting sustainable development agenda. Their principle is to strengthen the bottom-up approach to

Fig. 11.4 Green Incentives Award received during World Town Planning Day Selangor state level in 2019

Fig. 11.5 Platinum Winner Award (category Outdoor Public Park—waterfront 'blue park')

implement various activities under their plan towards a prosperous and sustainable future. Their plan is to include the involvement of federal and state agencies, private sectors and the community.

Kuala Langat District Council and Kg. Sijangkang community received Platinum Winner Award (for Outdoor Public Park-Waterfront blue park) for the My Place Award in 2019 (Fig. 11.5). The award was given by the Malaysian Institute of Planner (MIP), and the participation is from all over Malaysia. My Place Award celebrates community-led built environment projects that have transformed their locality. The Awards honour projects that have excelled in terms of design and community involvement/impact.[3]

[3] https://myplacescotland.org.uk/my-place-awards-entry/

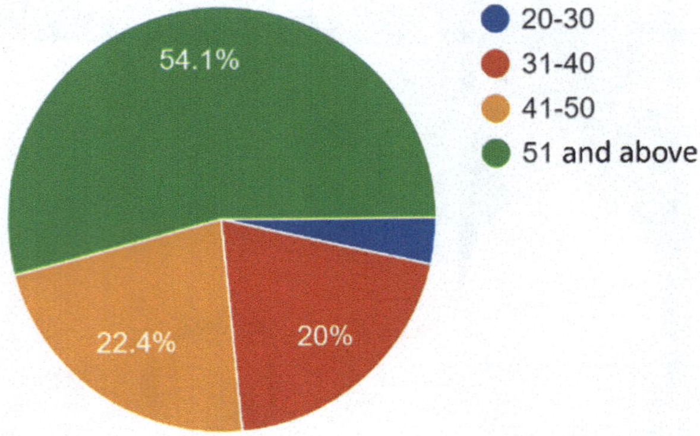

Fig. 11.6 The age group of the survey respondents

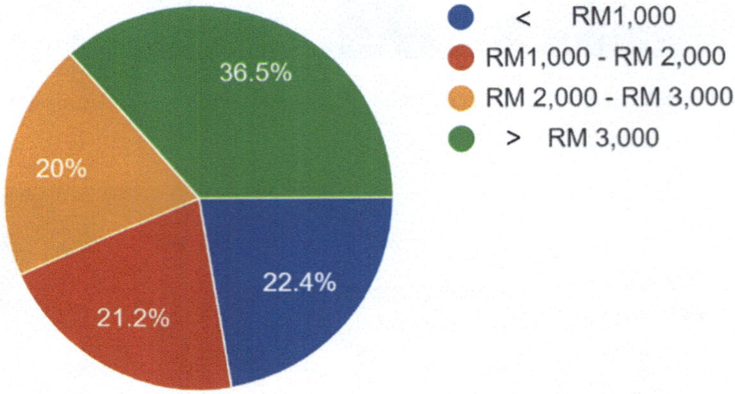

Fig. 11.7 The household income range of the survey respondents

11.3.4 Perceptions on Benefits of Mangrove Forests Conservation

The study involved virtual engagement with the community of Kg. Sijangkang to respond to the survey to identify their perception towards the benefits they gained from the Taman Rekreasi Paya Bakau Kampung Sijangkang. Figures 11.6, 11.7 and 11.8 show their demographic profile.

Figure 11.6 illustrates the age distribution, with 54% of the respondents aged 51 years old and above, 22% aged between 41 and 50 years old, 20% aged 31–40 years old and 4% aged 20–30 years old. This indicates that the senior group forms the active community members involved in the development and maintenance of Taman Rekreasi Paya Bakau Kampung Sijangkang. The senior-aged group have more flexible time to get involved in the community programs, aside from their other obligations. They are mostly pensioners and entrepreneurs.

Figure 11.7 shows that 37% of the respondents earned more than RM3000 a month, 22% earned below RM1000 a month, 21% earned between RM1000 and RM2000 and 20% earned RM2000–RM3000 a month. This implies that most

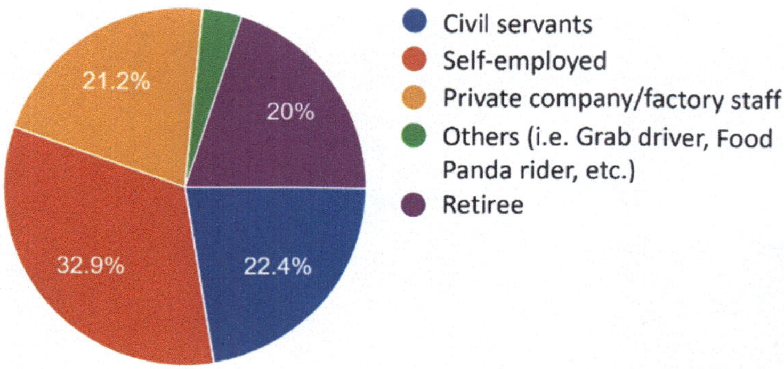

Fig. 11.8 The employment sectors of the survey respondents

respondents belong to the lower-income group of the bottom 40% (B40), where the average household income is below RM4850.00.[4] It shows the importance of sustaining the mangrove community program as it helps in generating and increasing their income to a comfortable level. The income generation and increment may not be direct, but the spill-off activities contributed to the overall economy of the community.

Referring to Fig. 11.8, 32% of the respondents are self-employed, 22% are civil servants, 21% are employed by a private company, 20% are retirees and 5% are gig workers. This indicates that the majority of the respondents are flexible to involve themselves and contribute to Taman Rekreasi Paya Bakau Kampung Sijangkang; they are among the self-employed and retiree groups. The results indicate that the spillover effects from the programme are important for the community well-being. Many of them are involved in food services activities that are highly dependent on the visitors to the park and event management in the area.

Figure 11.9 depicts the perceived benefits of the Taman Rekreasi Paya Bakau Kampung Sijangkang by the respondents. About 95% of the respondents perceived the park is preserving the environment of the mangrove trees, 89% on the protection from the riverside erosion, 84% on animal habitat protection in mangrove swamps and tourists attraction, 73% on reduction of the rate of river water pollution, 66% on cleanliness maintenance of the local area, 63% on helping the villagers to acquire new knowledge, 59% to help the economy of the villagers, 43% to help them communicate with local authorities and 37% to get media attention. These perceptions prove that the community who are actively involved in Taman Rekreasi Paya Bakau Kampung Sijangkang is well aware of the environment conservation effort and are willing to improve their socio-economic status and livelihoods.

[4] https://www.dosm.gov.my/v1/index.php?r=column/cthemeByCat&cat=430&bul_id=aVlJRDAvbjhWWEhQa1YvSWhsSjF3QT09&menu_id=L0pheU43NWJwRWVSZklWdzQ4TlhUUT09

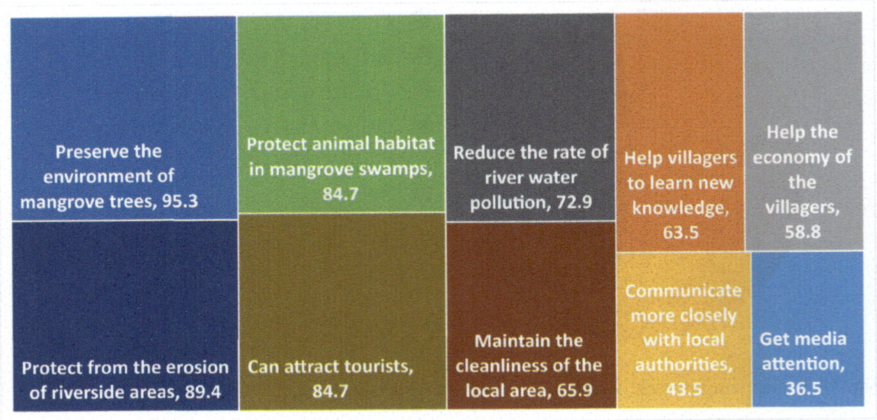

Fig. 11.9 Community perceptions on benefits of Taman Rekreasi Paya Bakau Kampung Sijangkang

As for attracting tourists, Fatimatuzzahroh et al. (2020) recognised social media as an unintended tool for promoting mangrove ecotourism, with visitors uploading media of mangrove tracks. In the case of the Kg. Sijangkang community, their Facebook page has become the primary platform for promoting and attracting visitors and tourists.

11.4 Conclusion

This chapter assessed the development of edutourism attraction by the local community of Kg. Sijangkang and analysed their edutourism programme of mangrove forests, how they started and their development process and success, and assessed the perceived benefits to the local communities. The interviews and survey showed the initial motivation of its establishment and the teamwork of all relevant stakeholders play a huge role and give impactful results.

Taman Rekreasi Paya Bakau Kampung Sijangkang establishment was mainly to clean the riverside area that has become a trash dumping site. Although faced with some resistance at the beginning of the establishment journey, the collaborations with local authorities and other agencies were soon developed, and the community started to accept and join the effort to develop and maintain the park and treat it as a communal project and activities.

The development process reflected by various activities and programmes were undertaken by the local community with various agencies, which included community engagement and replanting programmes, communal works and promotion on TV and radio. It shows the local community's commitment to working with agencies and societies from other areas to ensure the success and sustainability of the programme. The rewarding efforts lead to numerous achievements, including awards and recognition at the national and international levels, which served as honourable moments for the local community and further reinforce their motivations to continue conserving the mangrove forest as an edutourism park.

The local community highly perceived the benefits of developing the mangrove forest as the edutourism attraction helps preserve the environment and animals' habitat, offers protection from erosion and attracts tourists. These three benefits are the main drivers that keep the local community together in managing and sustaining the mangrove forest in Kg. Sijangkang.

The findings identified the importance of self-driven community and teamwork to progress and succeed in a conservation programme that could boost socio-economic status. The results also demonstrate the significance of collaborative efforts in the environmental conservation development programme.

Acknowledgements This study was inspired by the collaboration project between Taman Rekreasi Paya Bakau Kampung Sijangkang community and Institute of Tropical Forestry and Forest Products (INTROP), Universiti Putra Malaysia. The authors would like to express their sincere gratitude to Encik Suhaimi, Chairman of NGO managing Taman Rekreasi Paya Bakau Kampung Sijangkang, for his input and insights shared during this study. Special thanks to the active members of Taman Rekreasi Paya Bakau Kampung Sijangkang for their support and assistance.

References

Aheto, D. W., Kankam, S., Okyere, I., Mensah, E., Osman, A., Jonah, F. E., & Mensah, J. C. (2016). Community-based mangrove forest management: Implications for local livelihoods and coastal resource conservation along the Volta estuary catchment area of Ghana. *Ocean & Coastal Management, 127*, 43–54. https://doi.org/10.1016/j.ocecoaman.2016.04.006

Ardika, I. G. (2018). *Kepariwisataan Berkelanjutan; Rintis Jalan Lewat Komunitas*. Retrieved from https://gerai.kompas.id/belanja/buku/penerbit-buku-kompas/kepariwisataan-berkelanjutan/

Arrasyid, R., Urfan, F., , Ruhimat, M, Setiawan, I, Logayah DS (2021). Edutourism development model in UNESCO global geopark Ciletuh Palabuhanratu Sukabumi district. IOP Conference Series: Earth and Environmental Science 683,012120.

Bengtsson, M. (2016). How to plan and perform a qualitative study using content analysis. *NursingPlus Open, 2*, 8–14. https://doi.org/10.1016/j.npls.2016.01.001

Bourn, D., & Brown, K. (2011). *Young people and international development: Engagement and learning*. Development Education Research Centre (Research Paper 2). Institute of Education, University of London.

Budiharta, S., Meijaard, E., Wells, J. A., Abram, N. K., & Wilson, K. A. (2016). Enhancing feasibility: Incorporating a socio-ecological systems framework into restoration planning. *Environmental Science and Policy, 64*, 83–92.

Byrka, K., Hartig, T., & Kaiser, F. (2010). Environmental attitude as a mediator of the relationship between psychological restoration in nature and self-reported ecological behavior. *Psychological Reports, 107*, 847–859. https://doi.org/10.2466/07.PR0.107.6.847-859

Canty, S. W., Preziosi, R. F., & Rowntree, J. K. (2018). Dichotomy of mangrove management: A review of research and policy in the Mesoamerican reef region. *Ocean & Coastal Management, 157*, 40–49.

Crona, B., & Hubacek, K. (2010). The right connections: How do social networks lubricate the machinery of natural resource governance? *Ecology and Society, 15*(4), 5.

Darius, A., & Irena, S. K. (2018). Mangrove ecotourism development on Kakaralamo Island North Halmahera: Community perception, participation and development strategies. *IOP Conference Series: Earth Environmental Science, 175*(1), 012232. https://doi.org/10.1088/1755-1315/175/1/012232

Darsiharjo, S. U., & Saputra, I. M. (2016). Development of Ciletuh Geopark based on community participation as geotourism areas in Sukabumi District. *Resort and Leisure Management Journal, 13*(1), 55–60. https://doi.org/10.17509/jurel.v13i1.2036

Dewi, L. J. E. (2010). Searching for the shortest route in Bali using Dijkstra's algorithm. *National Seminar on Information Technology Applications, 1*, 46–49.

Duarte, C. M., Losada, I. J., Hendriks, I. E., Mazarrasa, I., & Marbà, N. (2013). The role of coastal plant communities for climate change mitigation and adaptation. *Nature Climate Change, 3*, 961.

Eddiwan, K. (2018). Success level of mangrove tree planting real in Kepulauan Meranti district, Riau, Indonesia. *Journal of Aquaculture & Marine Biology, 7*, 213–218. https://doi.org/10.15406/jamb.2018.07.00211

Fafurida, S. O., Sucihatiningsih, D. W. P., & Yoris, A. M. (2020). Sustainable strategy: Karimunjawa National Park Marine Ecotourism, Jepara, Indonesia. *International Journal of Scientific and Technology Research, 9*(3), 3234–3239.

Fatimatuzzahroh, F., Hadi, S. P., Purnaweni, H., & Sudarno. (2020). Does the intervention of regional authorities contribute to sustainable mangrove ecotourism? Case study on mangrove management at Karansong, West Java, Indonesia. *Ecological Questions, 31*, 7–14. https://doi.org/10.12775/EQ.2020.017

Giri, C., Ochieng, E., Tieszen, L. L., Zhu, Z., Singh, A., Loveland, T., Masek, J., & Duke, N. (2011). Status and distribution of mangrove forests of the world using earth observation satellite data. *Global Ecology and Biogeography, 20*(1), 154–159.

Gualard, F., Galánc, Á. L., Nelsond, E., & Soares-Filhoa, B. (2018). Perspective. Conservation lessons from Cuba: Connecting science and policy. *Biological Conservation, 217*, 280–288.

Huxham, M., Kimani, E., & Augley, E. (2004). Mangrove fish: A comparison of community structure between forested and cleared habitats. *Estuarine, Coastal and Shelf Science, 60*(4), 637–647. https://doi.org/10.1016/j.ecss.2004.03.003

Komiyama, A., Ong, J. E., & Poungparn, S. (2008). Allometry, biomass, and productivity of mangrove forests: A review. *Aquatic Botany, 89*(2), 128–137.

Kusmana, C., & Sukristijiono, S. (2016). Mangrove resource uses by local community in Indonesia. *Jurnal Pengelolaan Sumberdaya Alam dan Lingkungan (Journal of Natural Resources and Environmental Management), 6*(2), 217. https://doi.org/10.29244/jpsl.6.2.217

Le, H. D., Smith, C., Herbohn, J., & Harrison, S. (2012). More than just trees: Assessing reforestation success in tropical developing countries. *Journal of Rural Studies, 28*, 5–19. https://doi.org/10.1016/j.jrurstud.2011.07.006

Muhammad, F., Basuni, S., Munandar, A., & Purnomo, H. (2010). Kajian Daya Dukung Ekowisata Hutan Mangrove Blanakan, Subang, Jawa Barat Bioma Berk. *Bioma: Berkala Ilmiah Biologi, 14*, 64–72.

Nabila, A. R., & Yuniningsih, T. (2016). Analysis of community participation in the development of Kandri Tourism Village, Semarang City. *Journal of Public Policy and Management Review, 5*(3), 1–20. https://ejournal3.undip.ac.id/index.php/jppmr/article/view/12543

Nagelkerken, I., Blaber, S. J. M., & Somerfield, S. B. P. (2008). The habitat function of mangroves for terrestrial and marine fauna: A review. *Aquatic Botany, 89*(2), 155–185.

Ong, J. E., Gong, W. K., & Wong, C. H. (2004). Allometry and portioning of the mangrove *Rhizophora apiculate*. *Forest Ecology and Management, 188*(1), 395–408.

Polidoro, B. A., Carpenter, K. E., Collins, L., Duke, N. C., Ellison, A. M., et al. (2010). The loss of species: Mangrove extinction risk and geographic areas of global concern. *PLoS One, 5*(4), e10095. https://doi.org/10.1371/journal.pone.0010095

Putro, S. P., & Wasiq, J. (2015). Potensi Sumberdaya Alam Hayati Kawasan Mangrove Pasar Banggi Kabupaten Rembang Sebagai Objek Ekowisata. *Biosaintifika: Journal of Biology & Biology Education, 7*, 3955.

Rakhmanissazly, A., et al. (2018). Sustainable eco coastal development through corporate social responsibility (CSR) program. *IOP Conference Series: Earth and Environmental Science, 116*, 012038.

Renjaan, M. R., et al. (2020). *IOP conference series: Earth and environmental science*. 517012013.

Reimer, J. K. K., & Walter, P. (2013). How do you know it when you see it? Community-based ecotourism in the Cardamom Mountains of southwestern Cambodia. *Tourism Management, 34*, 122–132.

Sadono, R., Soeprijadi, D., Susanti, A., Matatula, J., Pujiono, E., Idris, F., & Wirabuana, P. (2020). Local indigenous strategy to rehabilitate and conserve mangrove ecosystem in the southeastern Gulf of Kupang, East Nusa Tenggara, Indonesia. *Biodiversitas Journal of Biological Diversity, 21*(3), 10353. https://doi.org/10.13057/biodiv/d210353

Singgalen, Y. A. (2020). Mangrove forest utilization for sustainable livelihood through community-based ecotourism in Kao Village of North Halmahera District. *Jurnal Manajemen Hutan Tropika, 26*(2), 155–168. https://doi.org/10.7226/jtfm.26.2.155

Spalding, M., Kainuma, M., & Collins, L. (2010). *World atlas of mangroves*. The International Society for Mangrove Ecosystems.

Thompson, B. S., & Friess, D. A. (2019). Stakeholder preferences for payments for ecosystem services (PES) versus other environmental management approaches for mangrove forests. *Journal of Environmental Management, 233*, 636–648.

Twilley, R. R., Chen, R. H., & Hargis, T. (1992). Carbon sinks in mangroves and their implications to carbon budget of tropical coastal ecosystems. *Water, Air, and Soil Pollution, 64*, 265–288. https://doi.org/10.1007/BF00477106

Walter, P. (2011). Gender analysis in community-based ecotourism. *Tourism Recreation Research, 36*(2), 159–168. https://doi.org/10.1080/02508281.2011.11081316

Wardhani, M. K., et al. (2020). The correlation of settlement preferences and community resilience post-disaster in Palu City. *IOP Conference Series: Earth and Environmental Science, 557*, 012050.

Weeks, R., Russ, G. R., Bucol, A. A., & Alcala, A. C. (2010). Shortcuts for marine conservation planning: The effectiveness of socioeconomic data surrogates. *Biological Conservation, 143*(5), 1236–1244. https://doi.org/10.1016/j.biocon.2010.02.031

Wylie, L., Sutton-Grier, A. E., & Moore, A. (2016). Keys to successful blue carbon projects: Lessons learned from global case studies. *Marine Policy, 65*, 76–84. https://doi.org/10.1016/j.marpol.2015.12.020

The Resilience of the Natural Resource Dependency of Indigenous People in a Wilderness Area: The Case of Virachey National Park, Cambodia

12

Phanith Chou and Hoeurn Cheb

Abstract

Virachey National Park has tremendous natural resources and is also home to populations of Indigenous and hill tribe people, with 85% of the park's population being Indigenous (Kavet and Brau). This chapter had two objectives: (1) examines whether the existing resource management practices of indigenous people represent resilience and (2) seeks to identify the most effective types of resource management approaches that should be promoted to ensure the resilience of Indigenous peoples' livelihoods. This study conducted surveys in six villages that began in 2015 using a structured questionnaire with 161 households to assess the natural resource dependency of Indigenous people. Additional surveys were conducted in 2019 and 2020 to collect further information through key informants, focus group discussions, a literature review and on-site observations to facilitate a deeper understanding of the current status of biodiversity and ecosystem services in Virachey National Park and identify the existing driving forces and pressures. In the study areas, 75% of the respondent households did fishing activities and 84% engaged in Non-Timber Forest Product (NTFP) collection, 48% in wildlife hunting and 15% in ecotourism, with average economic values of 1,095,834 (269.4 USD), 346,237 (85.1 USD), 371,057 (91.2 USD) and 568,125 riels (139.7 USD) per household per year, respectively. The social and economic benefits of ecosystem services in Virachey National

P. Chou (✉)
Department of Natural Resource Management and Development, Faculty of Development Studies, Royal University of Phnom Penh, Phnom Penh, Cambodia
e-mail: chou.phanith@rupp.edu.kh

H. Cheb
Department of Economic Development, Faculty of Development Studies, Royal University of Phnom Penh, Phnom Penh, Cambodia

Z. Samdin et al. (eds.), *Tropical Forest Ecosystem Services in Improving Livelihoods For Local Communities*, https://doi.org/10.1007/978-981-19-3342-4_12

Park are critically threatened by many driving forces, such as high timber demand, economic land concessions, climate change effects, decreased economic activities and pressures that include illegal logging, poaching, unsustainable NTFP extraction and farmland expansion. To counter these threats and increasing pressures, the Cambodian government, community protected areas and non-government organizations (NGOs) have organized strong efforts in response; yet it is not enough to address all the challenges noted. This research offers useful details and recommendations for researchers, policymakers, professionals and NGOs seeking insights into the state of Virachey National Park, its people, economic drivers and environmental threats.

Keywords

Forest resources · Driving force · Community resilience · Virachey National Park

12.1 Introduction

Cambodia boasts a vast area of tropical forests in mainland Southeast Asia, but its forests are under tremendous and alarming pressure (Chou, 2018; Cock, 2016; Milne & Mahanty, 2015). In the 1960s, 73% of Cambodia's land was forest (FA, 2010). Since this time, the Cambodian forest area has continuously declined to a current level of 47% of the country's total land area (MoE, 2020). This deforestation trend weakens ecosystem services and increases the vulnerability of forest-dependent communities, particularly Indigenous people (Chou, 2017; Milne & Mahanty, 2015; Watkins et al., 2016).

Deforestation in Cambodia has disrupted local populations' livelihoods, causing social injustice, cultural identity loss and regional conflicts (Milne & Mahanty, 2015). Forest-dependent people, comprising mostly Indigenous groups living in forested wilderness areas, are the most vulnerable, according to the Forestry Administration of Cambodia (FA, 2010). Indigenous communities suffer from land-use changes resulting from forest degradation, forest conversion into large-scale agro-industry sites, newly emerging settlements and infrastructure development. This situation reflects a lack of coherence in inclusive natural resource management to educate and regulate resource users' informal natural resource management procedures (Chou, 2019).

Subsequently, investigations into resilience decision making in collaborative natural resource use and management, and this concept referring to the capacities of local people and organizations to learn, adapt and individually and collectively manage the surrounding environment, according to Ross et al. (2009). Therefore, it is needed to understand whether Indigenous people in Cambodia have the capacity for such resilience. Virachey National Park was selected as a case study because it is an extremely controversial natural protected area facing critical uncertainty in land resource management. This chapter examines whether the existing resource management practices of indigenous people represent resilience and seeks to identify the

Table 12.1 Study site selection

Community protected area (CPA)	Village	Communes	District
O Tung	Lalai	Kok Lak	Veunsai
	Rak	Kok Lak	Veunsai
O Khampha	Rieng Vinh	Taveng Leu	Taveng
Mondul Yorn	Tabok	Taveng Kroam	Taveng
O Tabok	Tabok	Taveng Kroam	Taveng
Not yet a member of CPA	Tumpuong Reung Touch	Taveng Leu	Taveng
Not yet a member of CPA	Phyang	Taveng Leu	Taveng

Source: Created by the author based on an interview with the director of Virachey National Park

most effective types of resource management approaches that should be promoted to ensure the resilience of Indigenous peoples' livelihoods.

12.2 Methods

12.2.1 Site Selection

Virachey National Park was established as a Protected Area (PA) in Cambodia by Royal Decree on November 1, 1993, and was named an ASEAN Heritage Park in 2003. It is located in northeast Cambodia and covers 3325 km^2 (332,500 ha) in Ratanakiri and Stung Treng provinces. As the largest PA in Cambodia, it is one of the top priority areas for conservation in Cambodia and all of Southeast Asia. Virachey has been the home to Indigenous and hill tribe peoples for centuries and remains a centre of cultural diversity in Cambodia and the region (Ironside & Baird, 2003). Indigenous people (Kavet and Brau) make up 85% of the population in Virachey National Park (Baird & Dearden, 2003; Den & Sophark, 2012).

Virachey National Park has tremendous natural resources throughout its rich, semi-evergreen and deciduous forests; however, the park has lost a considerable amount of forest, particularly through the removal of valuable timber, land use and forest conversion that have accelerated over the past two decades. Hence, this study selected Virachey National Park due to its global biodiversity value and the increasingly urgent need for the intervention of resource management to properly regulate the use of the forest and the protection of its biodiversity and reduce further loss.

This study observed all four community protected areas (CPAs), which include O Tung, O Khampha, Mondul Yorn and O Tabok. Two villages that are not associated with CPAs were also selected to explore different characteristics of resource management, as seen in Table 12.1.

12.2.2 Data

This study conducted surveys that began in 2015 using a structured questionnaire including 161 households to assess the natural resource dependency of Indigenous people. Additional surveys were conducted in 2019 and 2020 to collect further information through key informants, focus group discussions, a literature review and on-site observations to facilitate a deeper understanding of the current status of biodiversity and ecosystem services in Virachey National Park and identify the existing driving forces and pressures. The interviews with key informants were conducted with chiefs, vice-chiefs, CPA residents, the Virachey National Park director, rangers and other NGO officers with experience in the region, including Fauna & Flora International (FFI), the United Nations Development Program, Bird Life and Conservation International. In addition, four focus group discussions were conducted that included a group of rangers from Veun Sai outpost, Taveng outpost, O Tong CPA and Mondol Yorn CPA. These group discussions explored the current state of natural resource management, capacities for additional natural resource management and the drivers of forest cover changes.

12.2.3 Analysis Framework

Driving Forces, Pressures, Stage, Impact, Response (DPSIR)
The construct of DPSIR was used to identify the current status of natural resources and biodiversity of Virachey National Park. The driving forces and pressures represent the factors affecting the change of resources together with its impact on Indigenous people's livelihoods. The capacities for a response from institutions and stakeholders were also examined to assess the potential resilience of Indigenous people through natural resource management.

Natural Resource Dependency
This study estimated the economic value of natural resources that Indigenous peoples directly gain from Virachey National Park, including Non-Timber Forest Product (NTFP) collection, fishing, wildlife hunting and ecotourism. The market approach was employed to estimate such direct use value because it considers both the benefits and costs of natural resource collection. In this study, we only observed labour costs because other costs of collection are difficult to obtain and were also minimal for the Indigenous peoples.

Institutional Framework Analysis
This study focuses only on the role of institutions in managing natural resources in a protected forest area. The institutions involved in natural resource management can be classified into external and local institutions to assess the roles, responsibilities, rights, power, linkages and learning processes of each stakeholder.

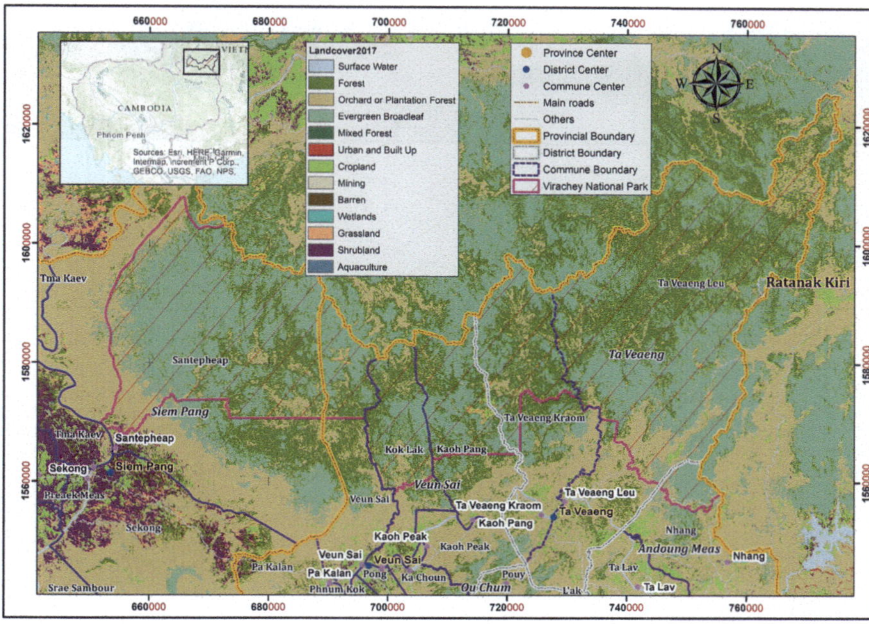

Fig. 12.1 Forest cover 2017 of Virachey national park. (Source: Created by the author based on Sevir Mekong regional land cover dataset)

12.3 Forest Resource Dependency of Indigenous People in Virachey National Park

12.3.1 Current Status of Forest Biodiversity and Ecosystem Services

Virachey National Park has unique terrestrial landscapes, including mountains, rivers, grasslands, forests and extensive biodiversity. It contains dense semi-evergreen forest, bamboo, mixed deciduous forest, montane forests, wildlife, waterfalls and mountains, as delineated in Fig. 12.1. According to forest cover 2017, evergreen forest, semi-evergreen forest, orchard or plantation forest and grassland covered 159,387 ha, 117,206 ha, 55,228 ha and 1459 ha respectively, and it also contains shrubland and other kinds of forests.

According to interviews, indigenous people rely on multiple ecosystem services. As seen in Fig. 12.2, forest biodiversity provides numerous provisioning services, including a variety of NTFPs, water for daily consumption, fish and bushmeat from wildlife. Details regarding the economic value of provisioning services are described below. Regarding regulating services, the terrestrial ecosystem of Virachey National Park contains a high potential for carbon storage of approximately 115,034,194

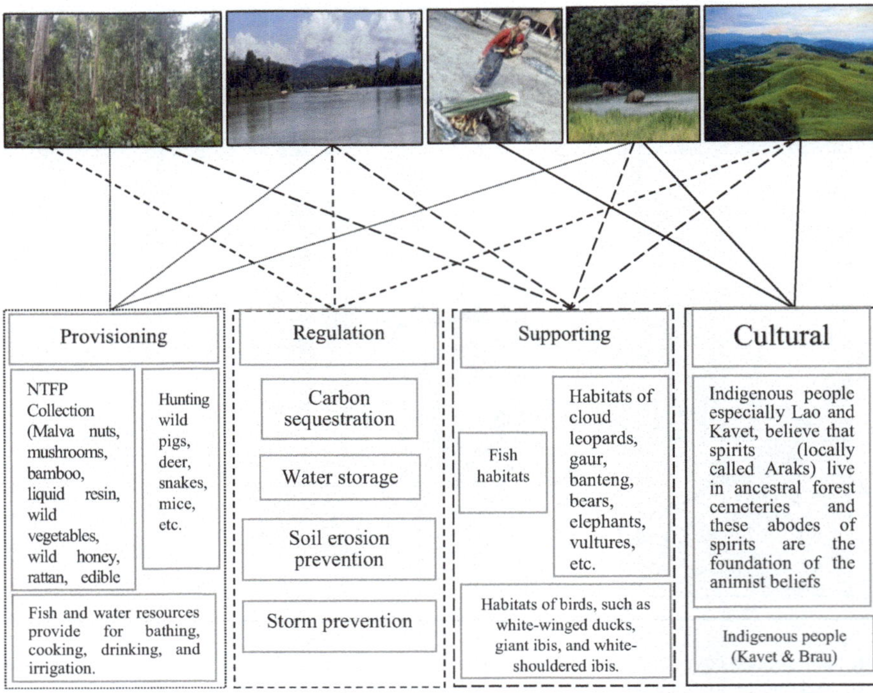

Fig. 12.2 Status of ecosystem services contribution to people of Virachey National Park. (Source: Created by the author based on interviews and site observations)

Megagramme Carbon (Mg C),[1] including 102,026,790 Mg C from forest cover and 13,007,404 Mg C from non-forest areas (Chou & Kiyoshi, 2021). The park is also believed to store water, reduce soil erosion and prevent storm degradation and human cost.

For cultural services, Virachey has been home to Indigenous and hill tribe people for centuries and remains a centre of cultural diversity in Cambodia and the region; 85% of the population is made up of Indigenous people (Kavet and Brau). Chamkar (agriculture), shifting/swidden ("slash-and-burn") cultivation and NTFP extraction are the main livelihood activities (Ironside & Baird, 2003). In terms of supporting services, Virachey's vast forests provide excellent habitat for multiple endangered species, including elephants, wild cattle and several spectacular, globally threatened primate and bird species (Fig. 12.2). According to the Virachey National Park Authority, the recorded wildlife includes at least 26 species of amphibians, 15 species of mammals, 100 species of birds, 35 species of reptiles, 37 species of fish, 19 katydid species and 30 ant species. The current flagship species in Virachey include the cloud leopard, gaur, banteng, bear, elephant and vulture.

[1] Megragramme Carbon (Mg C) is a tonne carbon.

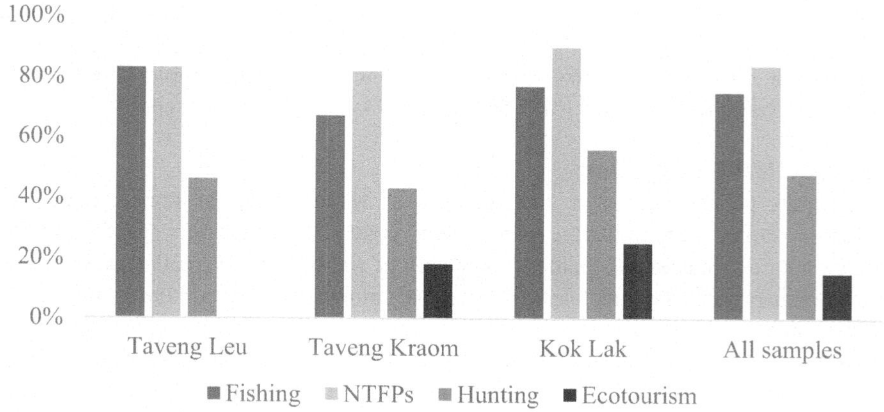

Fig. 12.3 Proportion of surveyed households benefiting from the selected natural resources. (Source: Household survey, 2015)

12.3.2 Forest Resource Dependency

The Indigenous peoples living in the study sites benefit from various activities that rely on natural resources. Only the selected direct use value from fishing, NTFP collection, wildlife hunting and ecotourism were analysed. As seen in Fig. 12.3, about 75% of the respondent households conducted fishing activities, 84% engaged in NTFP collection, 48% engaged in wildlife hunting, and 15% served as ecotourism guides; the average direct economic value of these selected resources was 1,095,834 (269.4 USD),[2] 346,237 (85.1 USD), 371,057 (91.2 USD) and 568,125 riels (139.7 USD) per household per year, respectively.

About 92% of the households surveyed were reliant on local natural resources as a source of livelihood. Within communes, 91% of the households in Taveng Leu, 90% in Taveng Kraom and 96% in Kok Lak accessed natural resources to support their livelihoods (Fig. 12.3). Annually, all the surveyed households earned 4,527,048 riels (112.8 USD) per household per year on average, and income from natural resources was an average of 1,491,856 riels (544.6 USD) per household per year. Hence, proceeds from natural resources contributed about 33% to annual household income.

12.3.2.1 Fishing

Fishing is one of the major sources of income that most Indigenous households rely on. The Sesan River is the source of fish resources. Among all samples, 75% of the respondent households generated an income from fishing. Half of them fished in the

[2] Official exchange in 2015 was 4068 Khmer riels/USD.

normal season, and 57% of them fished in the fishing season.[3] The average annual earnings from fishing were 1,664,198 riels (409.1 USD) per household, and the labour cost of fishing was 568,364 riels (139.7 USD) per household, making the economic value from fishing 1,095,834 riels (294.4 USD) per household.

12.3.2.2 NTFP Collection

Indigenous peoples collect a diverse array of NTFPs, such as honey, bamboo, bamboo shoots, rattan, wild cassava, wild potatoes, mushrooms, green leaves, malva nuts and medicinal plants. A majority of the households collected bamboo shoots, mushrooms and green leaves, whereas a minority collected honey, medicinal plants and other types of NTFPs. On average, 84% of the total respondent households generated income from NTFP collection, with an average economic value of 346,237 riels (85.1 USD) per household per year.

12.3.2.3 Hunting

Different types of hunting methods were used, with trapping as the most used method. Some households did not intend to hunt but used traps to protect crops from wild animals. According to the survey, hunted species included wild pigs, wild chickens, snakes, deer, mice, squirrels, weasels, ferrets, monitor lizards and birds. In sum, 48% of the total respondent households conducted hunting activities. The average economic value of the wild game was 371,057 riels (91.2 USD) per household per year. Including the labour cost, the benefit from the wild game was about 605,597 riels (148.9 USD) per year. This implies that to truly redirect these households from hunting, any alternative income-generating programme would need to provide at least 605,597 riels (148.9 USD) per year. In other words, the opportunity cost of ending hunting activities amounts to 605,597 riels (148.9 USD) per year per household. Some villages were aware that hunting did not present a good livelihood strategy, but the activity was engaged for survival (Box 12.1).

Box 12.1 Story of a Hunting Household

Living in Tumpuon Reung Touch village, Taveng Kraom commune, Mr. Dara[4] is 20 years old. He is currently living with five other family members involved in hunting to support their family. His household practices shifting cultivation, growing only rice. His family puts more effort into hunting and earns the highest income from wildlife hunting than other hunting households. When we asked them why they hunted, they said, "We know that hunting is not good for biodiversity. But so what? If we do not hunt, we do not have enough food to eat. Also, we can eat the meat from our hunting, also, we can sell it to local vendors so that we can get some more money to buy more food and support our family".

---

[3] The fishing season is the period between October and January (approximately 4 months) in which more fish are in the river than in other months, which are referred to as the normal season.

[4] Not his real name, to protect the anonymity of the respondent.

12.3.2.4 Ecotourism

The main ecotourism activity in Virachey National Park is trekking. There are three trekking routes in Ratanakari,[5] the Kalang Chhouy Sacred Mountain Trek, the O'Lapeung River Valley Trek and the Phnom Veal Thom Wilderness Trek. The Kalang Chhouy Sacred Mountain Trek and mountain biking are beside Kok Lak commune. The study villages in the commune benefited from this, with tourism activities taking place in the communities. Although the O'Lapeung River Valley Trek and kayaking are available in the Taveng Leu commune, only Yorn village in Taveng Leu benefited from these activities, but this village is difficult to access because it is far into the wilderness. Finally, the Phnom Veal Thom Wilderness Trek is in the Taveng District, and the study villages in Taveng Kraom commune benefited from this, although tourism activities do not take place directly in their communities. Only households offering local guides directly benefit from tourism.

Box 12.2 Story of the Household with the Highest Income from Ecotourism

Living in Rak village of Kok Lak commune, Mr. Samath[6] is 26 years old, and there are three people in his household. We asked Mr. Samath why his family does not hunt. He replied, "I think that if we can earn more money from other work such as being local guides, we do not need to go hunting because hunting is hard work. You have to walk far into the forest, and it is sometimes dangerous. For us, we want to stop hunting because if we hunt more wildlife, their numbers will decrease. Then tourists may not want to visit here".

The family was also asked how they could get more ecotourism work than other households. They laughed and said, "It depends on your connections with people in Banlong" (a town in Ratanakiri Province).

Around 15% of the respondent households benefited from ecotourism by working as local guides during the tourist season, but only 4% worked as local guides during the normal season. These households earned 635,208 riels (156.1 USD) per household, working for seven trips in the tourist season, receiving, on average, 93,542 riels (23.0 USD) per trip, each of which lasted about 3 days. Considering the labour cost, a household spent an average of 181,250 riels (44.6 USD) per household to support tourism activities. As a result, the economic value of tourism activities was only 453,958 riels (111.6 USD) per household in the tourist season. In contrast, the 4% of households that worked as local guides in the normal season earned an average of 650,000 riels (159.8 USD) per household, with an average of eight trips during this season and 3 days per trip. The average local guide fee was 85,000 riels (20.9 USD) per trip during the normal season, and the labour cost was 193,333 riels (47.5 USD) per season. Hence, the economic value of tourism in the normal season was 456,667

[5]New tourist sites may have been added since the time of this study.

[6]Not his real name, to protect the anonymity of the respondent.

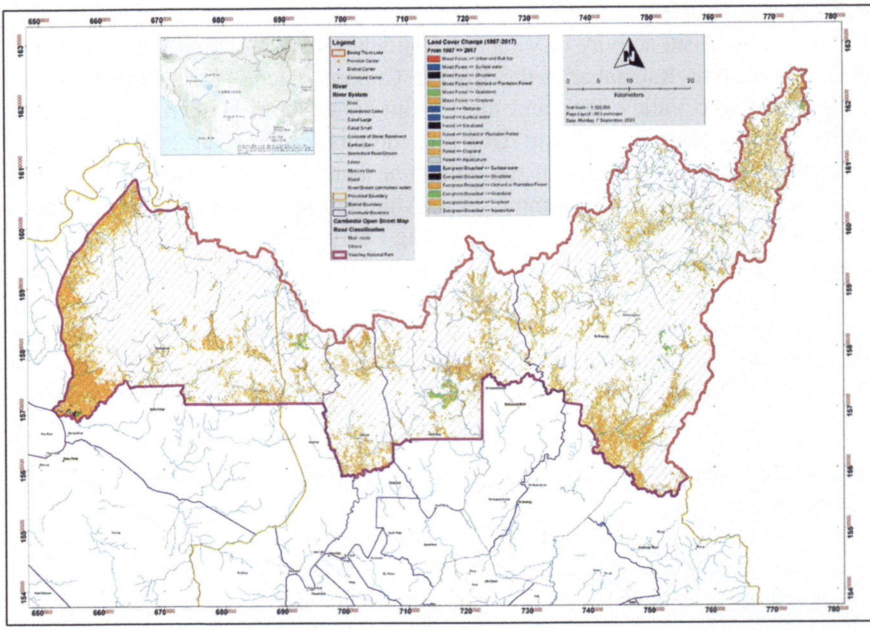

Fig. 12.4 Land use and land cover changes at Virachey National Park from 1987 to 2017. (Source: Created by the author based on Sevir Mekong regional land cover dataset. Note: Dark yellow identifies the critical land cover change area)

riels (112.3 USD) per household. In sum, the economic value of ecotourism was 568,125 riels (139.7 USD) per household per year for local guide households (Box 12.2).

12.4 The Driving Forces of Change in Forest Resources

Based on the land use and land cover analysis of satellite images from 1987 to 2017, the most recent data available reveals that human disturbance from logging and forestland conversion to agricultural plantations is a critical pressure.

Land-use changes took place rapidly along the border of Vietnam in the Dragon's Tail area and the Siem Pang district in Stung Treng province near the Lao border, as identified by the dark yellow spots in Fig. 12.4. According to the interviewees, the worst period of deforestation occurred from 1997 to 2007, when several economic land concessions were granted and through uncontrolled illegal logging. The forested had been converted to the plantation and deforested areas which impact critically to local communities and wildlife habitat.

The above-mentioned drivers lead to direct pressure on Virachey's wilderness and cultural landscape through activities that include logging, poaching, unsustainable NTFP extraction and agricultural land expansion, as seen in Fig. 12.4. These

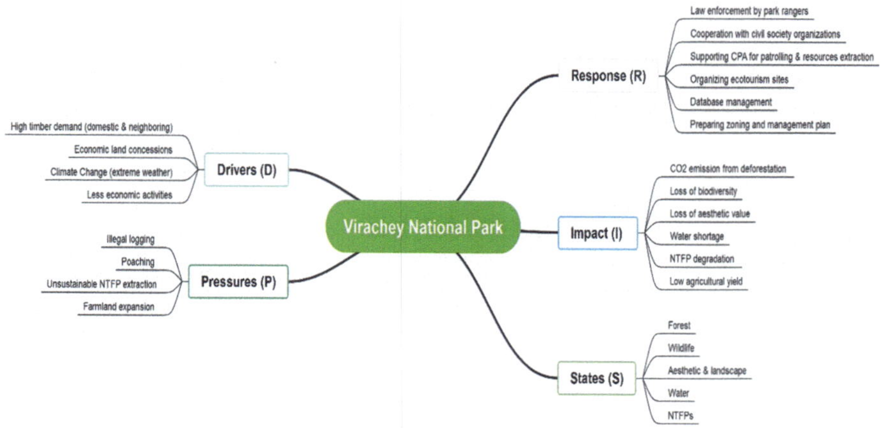

Fig. 12.5 Driving forces, pressures, states, impact and response at Virachey National Park. (Source: Created by the author based on interviews and literature)

pressures directly influence the ecosystem value of Virachey, including forests, wildlife, aesthetics and landscape, water and NTFPs. Indeed, numerous critical challenges are occurring simultaneously in Virachey, such as CO_2 emission from deforestation, loss of biodiversity, loss of aesthetic value, water supply shortages, NTFP degradation and decreased agriculture yield (Fig. 12.5).

To counter these threatening conditions, the Cambodian government at the national, provincial and park levels and communities, in collaboration with partner NGOs, have organized strong efforts in response to those drivers and pressures mentioned above, including increased law enforcement to crack down on illegal loggers, cooperation with development agencies (INGOs/NGOs), support for CPA patrolling and resources extraction, organizing ecotourism sites, development of adventure tour programmes and database management.

12.5 Co-management for Resilience in Natural Resource Management

The current institutional management of natural resources in Virachey National Park can be seen as a dynamic process of resilience in natural resource management by multi-stakeholders (Fig. 12.6). The Ministry of Environment and the Ratanakiri provincial department of the environment have rangers collaborating with the provincial administration to provide technical management practices and execute regulatory law enforcement to crack down on illegal logging, patrolling the forest and collecting biodiversity data using camera traps and smartphones. Rangers in Virachey also have a crucial role in collaborating with NGOs such as the FFI, the ASEAN Center for Biodiversity, Conservation International and Save Cambodia's Wildlife in conducting research, collecting biodiversity data, setting up camera traps

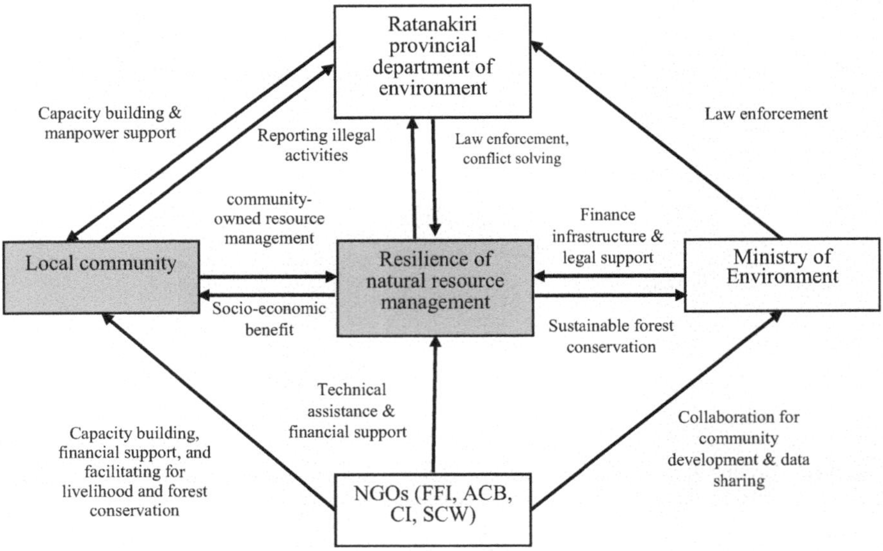

Fig. 12.6 Co-management process for natural resource management in Virachey National Park. (Source: Created by the author based on interviews and literature)

and contributing to the livelihood improvement of Indigenous people. NGOs also work directly with CPAs on capacity building, community organization, the establishment of ecotourism, setting up community forest patrol teams, developing strategies for sustainable NTFPs collection and devising community management plans. CPAs have a role in controlling natural forest areas under the common property right of the residents within its geographic boundary. Rangers and CPA members work as co-managers of the natural resources in Virachey National Park. Rangers control the resources at the core and conservation zones, while communities manage the resources within their community. Most often, both rangers and CPAs collaborated in reporting and enforcing laws on illegal poaching and logging activities.

Thus, the current natural resource management is referred to as co-management by Fabricius and Currie (2015) and Ross et al. (2009), representing the prospect of resilience for Indigenous people to collaboratively govern their natural resources with other stakeholders. As noted by Fabricius and Currie (2015), such co-management can be achieved during the short and medium term. There are some risks involved in the co-management approach for achieving resilience for Indigenous people in the long run, including uncertainty regarding institutional arrangements for management action, no sign of the driving forces of deforestation slowing down, lack of alternative sustainable economic activities to mitigate the natural resource dependency of Indigenous people, lack of technical assistance for sustainable natural resource management and transparent benefit sharing among the community and climate change-related effects.

The co-management process can be effectively implemented unless the local community has a strong capacity to interact with other stakeholders for deciding on natural resource management and benefit-sharing. NGOs, private sectors and provincial departments should engage the local community in the whole process of project design, project implementation and monitoring and evaluation. Instead of beneficiaries, the local community should be the project partner, so that they can learn and enhance their capacity towards the resilience of natural resource management in the long run.

12.6 Conclusion

Virachey National Park hosts a wealth of endangered ecosystems that are of immense importance to biodiversity and Indigenous communities. Indigenous people harness natural resources in the forest to support their livelihoods through fishing, NTFP collection, hunting and ecotourism. The social benefits deriving from ecosystem services in Virachey National Park have been critically threatened by numerous driving forces of high timber demand, economic land concessions, climate change, decreased economic activities and pressures such as illegal logging, poaching, unsustainable NTFP extraction and farmland expansion. Indeed, a multitude of critical challenges are occurring simultaneously in Virachey, such as CO_2 emission from deforestation, loss of biodiversity, loss of aesthetic value, water supply shortages, NTFP degradation and decreased agriculture yield.

To counter these circumstances, the Cambodian government at the national, provincial and park levels and communities, in collaboration with partner NGOs, have organized strong efforts to respond to these challenges by applying a co-management approach, with increased law enforcement to crack down on illegal loggers. CPA members have vital roles in managing natural resources for their sustainable livelihoods. Nevertheless, the current efforts are not enough to address all the challenges noted. Current conservation activities in Virachey cannot effectively save forests and biodiversity in the long run, and the forest is facing a turbulent future. Birge et al. (2016) advocated an "adaptive co-management" approach, which includes learning and structured decision-making processes for building resilience in the long term to advance Indigenous peoples' empowerment to govern their resources and ability to collaborate with other stakeholders to expand their role in all protected areas. Allen et al. (2011) proposed that adaptive management be developed to reduce the uncertain future through assessments and cross-scale trade-offs associated with structured decision-making (defining the problem, identifying the objectives, formulating evaluation criteria, estimating outcomes, evaluating trade-offs and making decisions) and learning processes (implement, monitor, evaluate and adjust). Regarding the case of the turbulent future of Virachey National Park, adaptive co-management should be demonstrated and adopted for achieving resilience of the natural resource dependency of Indigenous people.

Acknowledgements We thank Takahashi Industrial and Economic Research Foundation, ASEAN Centre for Biodiversity (ACB) and the Economy and Environment Program for Southeast Asia (EEPSEA) for supporting in funding the study. We thank the Ministry of Environment-Cambodia, Royal University of Phnom Penh, Aichi Gakuin University and Applied Social System Institute of Asia (ASSIA) at Nagoya University for their ongoing supports. We would like to express our sincere gratitude to Professor Kiyoshi Fujikawa, Dean of the Faculty of Economics, Aichi Gakuin University, for his enthusiastic encouragement and valuable advice. We also thank the editors and coordinator of this book for comments and editing our insights.

References

Allen, C. R., Fontaine, J. J., Pope, K. L., & Garmestani, A. S. (2011). Adaptive management for a turbulent future. *Journal of Environmental Management, 92*, 1339–1345. https://doi.org/10.1016/j.jenvman.2010.11.019

Baird, I. G., & Dearden, P. (2003). Biodiversity conservation and resource tenure regimes: A case study from Northeast Cambodia. *Environmental Management, 32*(5), 541–550.

Birge, H. E., Allen, C. R., Garmestani, A. S., & Pope, K. L. (2016). Adaptive management for ecosystem services. *Journal of Environmental Management, 183*, 343–352. https://doi.org/10.1016/j.jenvman.2016.07.054

Chou, P. (2017). The importance of non-timber forest products in rural livelihoods and ecosystem services at Phnom Prich Wildlife Sanctuary, Cambodia. *International Journal of Environmental and Sustainable Development, 8*(1), 7–13.

Chou, P. (2018). The role of non-timber forest products in creating incentives for forest conservation: A case study of Phnom Prich Wildlife Sanctuary, Cambodia. *Resources, 7*(3), 1–16. https://doi.org/10.3390/resources7030041

Chou, P. (2019). *The hidden value of non-timber forest products towards poverty alleviation and forest conservation: A case of Phnom Prich Wildlife Sanctuary, Mondulkiri Province, Cambodia*. Nagoya University.

Chou, P., & Kiyoshi, F. (2021). *Carbon storage and REDD+ in Virachey National Park*. Applied Social System Institute of Asia.

Cock, A. (2016). *Governing Cambodia's forests: The international politics of policy reform*. Nias Press.

Den, K. R., & Sophark, C. (2012). 15 Virachey National Park. In *Evidence-based conservation* (p. 216). Routledge.

FA. (2010). *Cambodia forestry outlook study*. Food and Agriculture Organization of the United Nations (FAO).

Fabricius, C., & Currie, B. (2015). Chapter 9: Adaptive co-management. In C. R. Allen & A. S. Garmestani (Eds.), *Adaptive management of social-ecological systems* (pp. 147–179). Springer.

Ironside, J., & Baird, I. G. (2003). *Wilderness and cultural landscape: Settlement, agriculture, and land and resource tenure in and adjacent to Virachey National Park, Northeast Cambodia*. Ministry of Environment.

Milne, S., & Mahanty, S. (2015). *Conservation and development in Cambodia: Exploring frontiers of change in nature, state and society*. Routledge.

MoE. (2020). *Cambodia forest cover 2018*. Ministry of Environment.

Ross, H., Grant, C., Robinson, C. J., Izurieta, A., Smyth, D., & Rist, P. (2009). Co-management and indigenous protected areas in Australia: Achievements and ways forward. *Australasian Journal of Environmental Management, 16*(4), 242–252. https://doi.org/10.1080/14486563.2009.9725240

Watkins, K., Sovann, C., Brander, L., Neth, B., Chou, P., Hoy, S., et al. (2016). *Mapping and valuing ecosystem services in Mondulkiri: Outcomes and recommendations for sustainable and inclusive land use planning in Cambodia*. WWF Cambodia.

Reconciling the Values and Needs of Wildlife and Local Communities: A Way Forward to Deal with Human–Wildlife Conflicts in Malaysia

13

Chong Leong Puan and Amera Natasha Mah Muhammad Adam Mah

Abstract

Human–wildlife conflicts represent a sign of undervaluing of various ecosystem services provided by wild animals. Such conflicts are common in many developing countries including Malaysia and occur when the values of wildlife have been overshadowed by short-term human gains as societies develop. This is indeed linked to market failure, i.e. wildlife and their habitats are often treated as public goods and hence they are prone to various threats and destruction. As a result, local communities have turned into victims of such conflicts affecting not only their long-term livelihood but also their lives. This chapter presents the various economic values of Malaysian wildlife and the consequences when these values are overlooked. Specifically, the associated costs due to local human–wildlife conflicts as well as conflict mitigation measures are discussed by highlighting local case studies. Besides local community involvement, the roles of various stakeholders including government agencies and non-governmental organisations are also included in the discussions. Finally, we review the current practices in Malaysia and recommend possible ways to reconcile the needs of local communities and wildlife to reduce these conflicts. In the end, this chapter intends to deliver a message that living harmoniously with wildlife is the way forward not

C. L. Puan (✉)
Faculty of Forestry and Environment, Universiti Putra Malaysia (UPM), Serdang, Selangor, Malaysia

Institute of Tropical Forestry and Forest Products (INTROP), Universiti Putra Malaysia (UPM), Serdang, Selangor, Malaysia
e-mail: chongleong@upm.edu.my

A. N. M. M. A. Mah
Faculty of Forestry and Environment, Universiti Putra Malaysia (UPM), Serdang, Selangor, Malaysia

Z. Samdin et al. (eds.), *Tropical Forest Ecosystem Services in Improving Livelihoods For Local Communities*, https://doi.org/10.1007/978-981-19-3342-4_13

only to sustain the livelihood of local communities in the developing countries but also the future development of these countries.

Keywords

Human–wildlife conflicts · Total economic value · Payments for ecosystem services · Stakeholder participation · Community engagement · Ecosystem functions · Malaysian wildlife

13.1 Being Rich: A Biological and an Economic Perspective

Located close to the equator and within the Sundaic region, Malaysia harbours one of the oldest rainforests on earth, which has also been recognised as a rich biome in which various flora and fauna have adapted to live. Blessed with the suitable climatic and edaphic factors, examples of terrestrial habitats in Malaysia include the mangrove, peat swamp, limestone, heath and riparian habitats. It is in such specialised habitats that unique species assemblages are found. Together with the marine habitats, the diverse ecosystems in Malaysia host more than 300 species of mammals, 780 birds, 240 frogs, 560 reptiles, 2060 marine and freshwater fishes and 13,000 butterflies and moths (MNRE, 2016). As with most tropical countries, intense species richness with large numbers of congeneric species, as well as many rare and endemic species, is a major characteristic of Malaysian biodiversity. With such high biodiversity, Malaysia has long been recognised as one of the 17 mega-diverse countries in the world based on the National Biodiversity Index (Groombridge, 1994) and part of the global biodiversity hotspots (Myers et al., 2000) and Global Ecoregions (Olson & Dinerstein, 2002). The inclusion of Malaysia in both the Endemic and Important Bird Areas (EBA & IBA; Yeap et al., 2007) further exemplifies the significance of natural habitats in the country.

In an ecosystem, different organisms proclaim their niche, for instance, by occupying different strata of the rainforests or different vegetation downstream or upstream of riparian habitat. The complex ecological interactions among these organisms and the interactions between biotic communities and abiotic elements (cycling of nutrients and transfer of energy) are what provide mankind with the various ecological functions and services. These floral and faunal resources sustain not only the livelihoods of the local communities but also the larger economy of tropical regions, providing tangible or intangible or use or non-use values to human societies. Traditionally, the indigenous people (e.g. the Orang Asli or Orang Asal in Malaysia) have been consuming forest products ranging from wild meat to non-timber products for food, clothing, building material, survival tools and medicines. To modern societies, wildlife provides important values as far as recreation, education and research are concerned.

13.2 Economic Growth and Human–Wildlife Conflicts

Being rich in natural resources inevitably implies a greater amount of resources available for exploitation. In Southeast Asia, the relative deforestation rate is the highest of any major tropical region. It is estimated that three-quarters of its original forests and up to 42% of its biodiversity will be lost by 2100 (Sodhi et al., 2004). Malaysia, as of 2021, is facing a biocapacity deficit of 71% implying the ecological footprint has exceeded the biocapacity of the country, estimated to be at 72 million gha (GFN, 2011). This reflects the great demand for natural resources and, consequently, pressure on native biodiversity in the country. At the moment, at least 50 mammals, 40 birds, 21 reptiles, 46 amphibians and 47 fish species in Malaysia are listed under one of the threatened categories in the IUCN Red List.

For many decades, people have been harvesting timber and non-timber forest products primarily for commercialisation purposes. It is primarily through logging activities that forests are degraded in terms of quantity and quality (Achard et al., 2002). Depending on the type of forest management practices, the associated effects may vary from the removal of large trees, destruction of forest structure and microhabitats, to land erosion. Such events alter tropical biotic communities both in terms of species and their numbers depending on the level of habitat disturbance (Berry et al., 2010). Overexploitation of forest may affect many interior forest specialists, and the recovery of the forest may be slow or may not reach its original condition as many forest-dependent species may have gone extinct (in the case of endemic species) locally. Similarly, building a dam may affect the aquatic ecosystems and disrupt the associated ecological interactions among aquatic organisms (Richter et al., 2010), which will eventually have a cascading effect on terrestrial organisms. It is through such activities and excessive use of natural resources that the growth of human populations is being supported, at the expense of biodiversity.

Throughout the world, wildlife and their habitats are being eliminated and disturbed at an alarming rate through deforestation, intensive agriculture, wetland drainage, invasive species, illegal poaching and pollution. In Malaysia, news about human–wildlife conflicts (HWC), roadkills and smuggling (Nijman, 2010) are making headlines more often than before. Such incidents are often complicated and interrelated with issues such as habitat encroachment, poaching, loss of human lives, damage to properties and the livelihoods of indigenous people (Clements et al., 2010; Saaban et al., 2011).

Out of 86,040 HWC complaints (e.g. wildlife disturbance and attacks as well as damages and losses of livestock and properties due to wild animals) recorded by the Department of Wildlife and National Parks (DWNP) from 2006 to 2015, most were associated with the long-tailed macaque (*Macaca fascicularis fascicularis*) (66%), followed by the Asian elephant (*Elephas maximus*) (9%), the wild boar (*Sus scrofa*) (7%), the common palm civet (*Paradoxurus hermaphroditus*) (6%) and the pig-tailed macaque (*Macaca nemestrina*) (3%). From 2004 to 2015, a total of 781 cases involving human death and injuries were recorded involving snakes (505), wild boars (99), primates (92), elephants, bees/hornets (29), Malayan sun bear (*Helarctos malayanus*) (7) and Malayan tiger (*Panthera tigris jacksoni*) (9) (Saaban et al., 2016). Conflict undeniably causes fear and negative perception among the local communities towards wildlife. In terms of monetary value, a total of RM550,233 was lost due to livestock depredation between 2007 and 2015 with the highest loss reported from Terengganu (RM284,240) followed by Negeri Sembilan (RM77,170) and Pahang (RM58,524) (Saaban et al., 2016). In Sarawak, cases mainly involved the saltwater crocodile (*Crocodylus porosus*) and the long-tailed macaque in 2013–2015, i.e. 28 of 110 crocodile attacks resulted in 15 deaths and 13 severely injured, whereas it was 56 cases for macaques (Tisen, 2016).

Roadkill is another form of HWC involving many wildlife species (e.g. civets, pangolins, monitor lizards and snakes) (DTCP, 2009; Saaban et al., 2016). From 2011 to 2019, a total of 3386 cases of roadkill were recorded in Peninsular Malaysia with the highest cases recorded from the state of Johor (702 cases) (Ten et al., 2021). Based on 2010–2014 records, three recorded species of roadkill were listed as Endangered (Malayan tapir *Tapirus indicus*, Asian wild dog *Cuon alpinus* and Asian elephant). Three recorded species of roadkill were listed as vulnerable (smooth-coated otter *Lutrogale perspicillata*, pig-tailed macaque and binturong *Arctictis binturong*). The majority of road mortality occurred on oil palm plantation roads compared to highways (Jamhuri et al., 2020). Specifically, mortality of small- and medium-sized mammals due to roadkill was greater at locations closer to fragmented compared to continuous forests (Jamhuri et al., 2020). Such scientific-based findings related to roadkills would contribute towards better land use management and decision-making to reduce HWC occurrences. For instance, this can be done by identifying hotspots of roadkills and avoiding building roads that cut across contiguous forest habitats as well as creating wildlife corridors that connect forest fragments.

13.3 Market Failure and Wildlife Overexploitation

Knowing the consequences of HWC, one may raise questions on why development projects are resulting in losses rather than gains. In reality, most of the environmental problems including HWC occurring nowadays are directly and indirectly linked to human activities, more so when it involves unsustainable decisions to make trade-offs between preserving natural resources and development—we tend to underestimate or overlook the value of many ‘unpriced’ natural resources, including wildlife. On the other hand, the benefits of human activities are often being overestimated. As a result, the related, often adverse, impacts of development have to be borne disproportionately by other parties (Dixon & Sherman, 1991). For instance, a development project would encroach an existing habitat of elephants, forcing them out and to wander into nearby plantations, thereby threatening the lives of the local communities and their properties (Saaban et al., 2011). Such threats are usually not properly accounted for when a land conversion project is being implemented, and the wandering elephants would then be blamed for any accidents that happen. This shows that there is still a lack of understanding among the general public when it comes to sources of HWC, as well as the value or importance of having these large herbivores (Campos-Arceiz et al., 2012) in an ecosystem.

The benefits of natural resources are undervalued for several reasons, which can be described as market failure or the failure of human societies to allocate resources efficiently. Natural resources are hard to express in monetary terms such as the ecological services and processes that sustain biodiversity. Unlike typically marketed goods, most natural resources are also regarded as common property implying that no one owns them. Wildlife that is not already traded in a marketplace may often be regarded as free goods, making them vulnerable to various threats as protection measures promoted by ownership cannot be adopted. The same goes for natural habitats that are needed to support wildlife. With only a few protected areas privately owned and managed, many natural areas are generally regarded as 'open access' (Ooi, 1990) for uncontrolled use and exploitation. In the case of natural areas that are established for recreational purposes, users may enjoy visiting, but unless a fee is imposed, there is no valuation on how much such 'enjoyment' is worth. Hence, regardless of whether the areas are under protection, under normal market mechanisms, the desired land area to be protected remains unknown as no individual is required to pay for its protection (Dixon & Sherman, 1991). Furthermore, since most protected areas are established through public investments made by governments, undervaluation or failure to estimate monetary benefits may provide little motivation concerning the allocation of funds needed for proper management to derive continued provision of the benefits associated with these areas.

The nature of some natural resources can further impede their protection. Unlike minerals and crops, some resources may be almost impossible to achieve exclusion, either technologically or physically. For instance, migratory birds on land and fish in the ocean can travel over large distances across manmade borders, and therefore protection of these species can be very difficult. The same goes for ecological processes, such as watershed protection and carbon sinks, which are dispersed over the landscape at large and are not confined to an area. Consequently, an efficient allocation and protection of resources may be problematic without some forms of government intervention, for instance, employing law enforcement. In Malaysia, the Wildlife Conservation Act 2010 [Act 716], Access to Biological Resources and Benefit Sharing Act 2017 [Act 795], International Trade in Endangered Species Act 2008 [Act 686], National Parks Act 1980 [Act 226] and Fisheries Act 1985 [Act 317] have been enacted to serve such a purpose.

For various reasons mentioned above, under typical market mechanisms, the economic values or benefits we derive from natural resources are not entirely accounted for, but the costs of their protection are often being mentioned. Moreover, there is a tendency to understate the risks involved when the proper functioning of ecosystems is affected due to exploitation and destruction of natural resources. For example, overfishing distorts the entire food chain in the ocean altering marine ecosystem structure and functions, leading to serious social, economic and environmental implications (Ooi, 1990). The largest marine turtle in the world, the leatherback sea turtle (*Dermochelys coriacea*), used to nest in great numbers on the east coast of Peninsular Malaysia. However, the over-harvesting of its eggs caused a dramatic population decline from more than 10,000 nests in the 1950s to none today (Chan, 2006; Chan & Liew, 1996). The local extinction of the species in Malaysia has also affected tourism (i.e. from turtle watching; MNRE, 2016) in the area. With an increase in pressure on natural resources from economic development, giving value to natural resources in a more holistic approach is one way towards improving policy and decision-making (Davison, 1997) to encourage the use of natural resources in a more sustainable manner.

13.4 Valuing Wildlife Through Various Approaches

Today, we choose to pay for goods and services according to our choice or priority which reflects the value that we put on the things we buy. Conceptually, the value of unpriced natural resources such as wildlife is analogous to the value of goods and services that we are willing to make trade-off or give up in exchange for getting those benefits associated with the resources. As such, by denominating it a monetary unit, wildlife resources can be valued for a variety of uses and reasons. The concept of total economic value (TEV; Fig. 13.1) is often used to categorise the values of natural resources. The TEV framework comprises two broad categories, i.e. use and non-use values. Total use value comprises the actual use value which is divided into current use value and option value. These values may come in the form of direct or indirect use values. For wildlife, the direct use values may be derived from harvesting bushmeat or wild honey or taking wild animals legally as pets, which can also be referred to as consumptive use values. Globally, the annual retail value of the live reef fish trade was estimated at US$1.2 billion (about RM4.8 billion) in 1995, of which US$1.0 billion came from live food fish while US$200 million was from aquarium fish (Barber & Pratt, 1998). Ng and Tan (1997) further estimated that the aquarium fish trade in Southeast Asia can reach S$100–200 million (about RM290–590 million) annually. In the case of edible bird's nest, the global trade was estimated to be over US$1.6 billion (about RM6.5 billion), and the annual production in Malaysia alone has reached RM1 billion (Runckel, 2010; Thorburn, 2015). These are examples of consumptive use-values.

On the other hand, the non-consumptive use values are obtained from wildlife-based recreation or ecotourism activities such as birdwatching and nature photography. With the biodiversity in Malaysia, wildlife has certainly become a prime attraction concerning the nation's ecotourism industry. Examples of famous ecotourism activities include watching of vertebrates [e.g. the Bornean orangutan (*Pongo pygmaeus*) at the Sepilok Orangutan Rehabilitation Centre and the sun bear at the Bornean Sun Bear Conservation Centre (BSBCC) in Sabah; the Asian elephant at the National Elephant Conservation Centre, Kuala Gandah, Pahang (Kaffashi et al., 2015); migratory raptors at Tanjung Tuan, Melaka (Puan et al., 2014); the kelah or red mahseer (*Tor tambroides*) at Lubuk Tenor, Taman Negara Pahang] as well as activities involving invertebrates [e.g. the fireflies at Kampung Kuantan Firefly Park in Kuala Selangor, Selangor (Mohd Shahwahid et al., 2013); the Rajah Brooke's birdwing (*Trogonoptera brookiana*) in Ulu Geroh, Perak; and squid jigging in Kuala

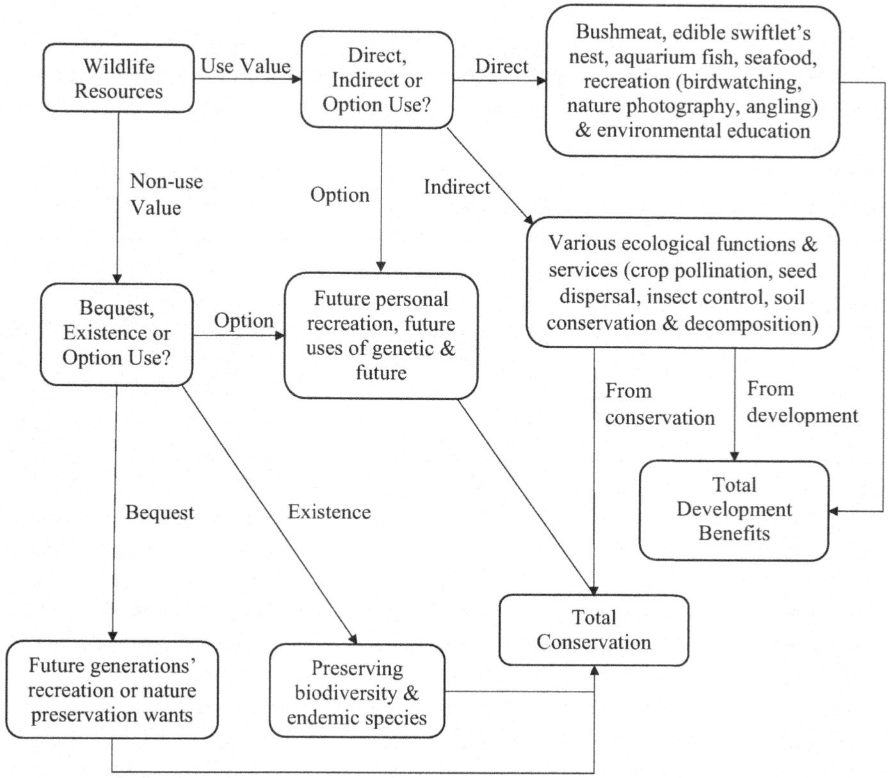

Fig. 13.1 The total economic value of wildlife (Bateman & Turner, 1995)

Terengganu, Terengganu]. The economic values of many of such ecotourism spots in Malaysia have yet to be estimated.

The indirect use values are values gained from indirect uses of the resources. For wildlife, indirect use values include various ecological roles of wildlife in maintaining ecosystem functions and services that are beneficial to humans. For instance, bats have been known to be a crucial pollinator and seed disperser in an ecosystem (Kingston, 2010). Owing to their ecological roles, disturbance to bat populations will affect forest regeneration as well as the yield of many fruit crops (Fujita & Tuttle, 1991) including durian (*Durio zibethinus*) and petai (*Parkia speciosa*) of high economic return. As of 2019, durian export in Malaysia was reported to be US$22.3 million (Safari et al., 2021) implying the significant indirect use value we gained from ecological services provided by bats. Another example of a nocturnal animal contributing to such a value is the barn owl (*Tyto alba javanica*). In Malaysia, the owl species has been successfully introduced in agricultural areas, mostly in oil palm plantations (Puan, 2013) due to the abundance of rodents as food and the provision of nest boxes in these plantations. A study on the predation efficiency of the owl showed that a breeding pair and their chicks could consume up to 1200–1500 rodents annually (Duckett & Karuppiah, 1990). In most plantations, the owls provide supplemental biological control of rodent pests, which has helped to reduce the baiting cost of RM2–RM30 per ha annually (Hafidzi & Saayon, 2001). These are examples of the indirect use values we obtain from wildlife.

Similar to the concept of paying for insurance coverage, option value represents the value derived from knowing the existence of an opportunity or leaving an option of using certain resources in the future. For example, people would be willing to pay for the preservation of the leatherback sea turtle (*Dermochelys coriacea*) to have an opportunity to observe the animals in the future. On the other hand, bequest value is the benefit obtained by individuals from knowing that their future generations will be benefited in the same way in the future. This is evidenced by the contribution of money and energy by volunteers in environmental programmes to create awareness of preserving natural resources for future generations. Similarly, existence value is the benefit derived from simply knowing that the resources exist even when one may not have utilised or had any intention of using these resources. Existence value can be significant especially when it involves threatened species like hornbills or very unique habitats like the montane forest. A person may be willing to contribute a certain amount of money to protect these resources, although he or she is not a birdwatcher or has limited knowledge of birds. Over the last two decades, many environmental NGOs, projects or programmes have been established in Malaysia of which their major intention can be directly and indirectly linked to indirect use, bequest and/or existence values. With each targeting on a specific taxon, such programmes or projects include the Sea Turtle Research Unit (SEATRU) formed in 1998 in Terengganu; the Raptor Count that started in 2000 at Tanjung Tuan, Melaka; the Malaysian Conservation Alliance for Tigers (MYCAT) founded in 2003; the Save Our Seahorses (SOS) that commenced in 2005; the Hornbill Volunteer Programme that started since 2008 at the Belum-Temengor Forest Complex; the Turtle Conservation Society of Malaysia that established in 2011; and the Sabah Shark Protection Association formed in 2015.

SEATRU has been operating for more than two decades with its volunteer programme running annually from April to September. Over the years, the responses of volunteers have been overwhelming. A willingness of these volunteers to pay a participation fee (i.e. RM500–800 for each local participant) can indirectly reflect the value people place on the conservation of wildlife, to a certain extent, and in this case, the sea turtle. In addition, many of such organisations or programmes also receive a donation from public or private funders to carry out their work, e.g. RM11.4 million being donated to the Borneo Rhino Alliance (BORA) in 2009 for saving the then critically endangered Sumatran rhinoceros (*Dicerorhinus sumatrensis*), which unfortunately has now gone extinct locally. In Malaysia, as well as Southeast Asia in general, more and more research units or programmes have been formed, most of which offer internship and/or volunteering and outreach programmes. Examples include the Tropical Research and Conservation Centre (TRACC) in Borneo initiated in 2001; the Southeast Asian Bat Conservation Research Unit (SEABCRU) founded in 2007; Rimba formed in 2010; the

Management and Ecology of Malaysian Elephants (MEME) founded in 2011; the Hose's Civet and Small Carnivore Project in Borneo (HOSCAP Borneo) commenced in 2012; and MareCet Research Organization established in 2012.

In Malaysia, other than using market price, the contingent valuation method (CVM; e.g. Amiry et al., 2009; Mohd Rusli et al., 2009; Puan & Zakaria, 2006; Siew et al., 2015) and travel cost method (TCM; e.g. Syamsul Herman et al., 2013) are two valuation methods that have frequently been applied in estimating the value of wildlife (particularly non-traded animals) and/or wildlife-based activities. CVM requires people to state their value directly through a survey or interview about their willingness to pay (WTP) for a particular ecosystem service. On the other hand, as its name implies, TCM calculates the related value based on the amount of money and time to access and engage in certain activities (mainly recreational activities). For example, visitors to the Panti Forest Reserve in Johor are required to pay RM150 to get a permit to access the forest, a popular site for birdwatching. Such payment can then be used as a starting bid to measure the economic value of the forest via the concept of WTP, in addition to the associated costs to make such travel.

Fraser's Hill in Pahang is another renowned birdwatching site (Noramly & Yeap, 2001; Puan & Zakaria, 2006), which attracts both local and foreign birdwatchers. The money and time that these birdwatchers are willing to spend are an indication of the value they placed to engage in wildlife-based activities at the site. However, does this mean Fraser's Hill has no value to non-birdwatchers or people who have never visited the site? Puan and Zakaria (2006) attempted to answer this question by conducting a questionnaire survey of the WTP (Bateman & Turner, 1995) of both birdwatchers and non-birdwatchers for the protection of Fraser's Hill as a bird sanctuary. It was estimated that respondents were willing to contribute at least RM30.40 per year for that purpose. By multiplying this estimate by the annual number of visitors, the conservation value or benefit of protecting Fraser's Hill was estimated at RM1,463,760 per year.

Another example of a wildlife value estimation approach would be based on the costs of damage avoided, replacement or substitution, e.g. the value of the endangered milky stork (*Mycteria cinerea*) in Malaysia was about RM404,000 based on the costs of species reintroduction in 1997 (Davison, 1997). To deal with HWC in Kelantan, some local villagers took the initiative by installing fences (58.82%) or shooting wildlife (41.8%), while others chose to ignore the conflict (31.02%) or made a report to the related authorities (29.95%). Some even tried to approach poachers to resolve conflicts as it was perceived to be more effective (Hassan et al., 2017). All these actions have their respective costs and challenges that the villagers have to bear (Nyhus et al., 2005). On a larger scale, the Malaysian government had allocated RM2.5 million to purchase land for the building of wildlife corridors, habitat restoration and HWC monitoring (Ten et al., 2021) as part of the Central Forest Spine (CFS) Master Plan for Ecological Linkages. The CFS aims to restore connectivity among forest complexes in Peninsular Malaysia (DTCP, 2009; Saaban et al., 2016) as a long-term solution to deal with HWC. Other efforts include having restrictions on the expansion of lanes on highways, monitoring viaduct effectiveness to habitat connectivity and stepping up anti-poaching efforts (Wong et al., 2018) by involving the local community (Hassan et al., 2017). Viaduct usage is also being improved via habitat enrichment, e.g. by deploying artificial salt licks, establishing pastures and planting local fruit trees (Zainol et al., 2021). All these measures exist and therefore should be properly accounted for (e.g. cost-benefit framework) in development projects.

13.5 The Way Forward: Wildlife Value Capturing and Inclusion

Malaysia harbours rich and diverse wildlife resources, and consumers or stakeholders may place very different economic values on these resources. From a Malaysian perspective and based on mostly local case studies, this chapter highlighted not only the current environmental issues that have resulted from market failure as well as ignoring the TEV of wildlife resources in the country but also the potentials of capturing these values in the country. Furthermore, this chapter also reveals that there are plenty of research opportunities to estimate the economic values of wildlife in Malaysia.

With the country's population of more than 31 million, greater pressure will continue to befall wildlife and their habitats in Malaysia over time. Even with the current Wildlife Conservation Act 2010 that has had the majority of wildlife species listed as totally protected or protected, this still does not guarantee the long-term protection of these animals. Under limited public budgets, wildlife conservation will need to compete with many other social goals, whose related costs and benefits are already clear in monetary terms. In other words, if wildlife resources are merely being treated as raw materials to be exploited, no amount of scientific knowledge or justification will solve the problem of wildlife loss. With such a condition, more wildlife is expected to follow the footsteps of the already locally extirpated Javan rhinoceros (*Rhinoceros sondaicus*), green peafowl (*Pavo muticus*), leatherback sea turtle and Sumatran rhinoceros.

Considering the increasing risks of losing our natural resources and biodiversity and having more HWC in the future, developing countries must find ways to reconcile the need for wildlife conservation and economic development, as well as bridging the gap between policymakers and conservationists via stakeholder participation, community engagement and integration of scientific-based knowledge. In addition, this can also be achieved through integrating economic values of wildlife resources and mainstreaming the importance of such values into all levels of decision- and policy-making processes to prevent undesirable externalities, as mentioned above. All in all, the TEV of wildlife needs to be sufficiently identified and empirically assessed and captured through appropriate economic instruments (Bateman & Turner, 1995; Davison, 1997). All these are requirements that will facilitate the implementation of Payments for Ecosystem Services (PES; MNRE, 2016; Wunder, 2005) in the future. Capturing these values would provide economic incentives coupled with finding a sustainable financing mechanism to safeguard our wildlife resources and mitigate HWC so that humans and wildlife can co-exist harmoniously.

Acknowledgements We are grateful to Wei Lun Ng, Su Ping Ong and Kelvin Peh for their constructive comments on the earlier versions of this manuscript. We thank the Laboratory of Bioresource Management, Institute of Tropical Forestry and Forest Products (INTROP), Universiti Putra Malaysia, for the opportunity and support to publish this book chapter.

References

Achard, F., Eva, H. D., Stibig, H. J., et al. (2002). Determination of deforestation rates of the world's humid tropical forests. *Science, 297*, 999–1002.

Amiry, S. A., Mohd Rusli, Y., Radam, A., et al. (2009). Recreational demand in bird sanctuary: The case of Kapar Bird Sanctuary, Kelang, Malaysia. *International Journal of Business and Management, 4*(12), 99–111.

Barber, C. V., & Pratt, V. R. (1998). Poison and profits: Cyanide fishing in the Indo-Pacific. *Environment: Science and Policy for Sustainable Development, 40*(8), 5–34.

Bateman, I. J., & Turner, R. K. (1995). Valuation of the environment, methods and techniques: The contingent valuation method. In R. K. Turner (Ed.), *Sustainable environmental economics and management: Principles and practice* (pp. 120–191). Wiley.

Berry, N. J., Phillips, O. L., Lewis, S. L., et al. (2010). The high value of logged tropical forests: Lessons from northern Borneo. *Biodiversity and Conservation, 19*, 985–997.

Campos-Arceiz, A., Traeholt, C., Jaffar, R., et al. (2012). Asian tapirs are no elephants when it comes to seed dispersal. *Biotropica, 44*(2), 220–227.

Chan, E. H. (2006). Marine turtles in Malaysia: On the verge of extinction? *Aquatic Ecosystem Health & Management, 9*(2), 175–184.

Chan, E. H., & Liew, H. C. (1996). Decline of the leatherback population in Terengganu, Malaysia, 1956–1995. *Chelonian Conservation and Biology, 2*(2), 196–203.

Clements, R., Rayan, D. M., Zafir, A. W. A., et al. (2010). Trio under threat: Can we secure the future of rhinos, elephants and tigers in Malaysia? *Biodiversity and Conservation, 19*, 1115–1136.

Davison, G. W. H. (1997). *Economic valuations and costings of animal diversity*. WWF Malaysia.

Dixon, J. A., & Sherman, P. B. (1991). *Economics of protected areas: A new look at benefits and costs*. Earthscan Publications Ltd.

DTCP. (2009). *Final report CFS I: Master plan for ecological linkages: May 2009*. Department of Town and Country Planning.

Duckett, J. E., & Karuppiah, S. (1990). A guide to the planter in utilizing Barn Owls (*Tyto alba*) as an effective biological control of rats in mature oil palm plantations. In B. S. Jalani (Ed.), *Proceedings of the 1989 International Palm Oil Development Conference: Agriculture* (pp. 357–372). Palm Oil Research Institute of Malaysia.

Fujita, M. S., & Tuttle, M. D. (1991). Flying foxes (Chiroptera: Pteropodidae): Threatened animals of key ecological and economic importance. *Conservation Biology, 5*(4), 455–463.

GFN. (2011). *Global footprint network: Advancing the science of sustainability*. Retrieved August 26, 2021, from www.footprintnetwork.org

Groombridge, B. (Ed.). (1994). *Biodiversity data sourcebook*. World Conservation Monitoring Centre (Comp.), World Conservation Press.

Hafidzi, M. N., & Saayon, M. K. (2001). Status of rat infestation and recent control strategies in oil palm plantation in Peninsular Malaysia. *Pertanika Journal of Tropical Agricultural Science, 24*, 109–114.

Hassan, S., Hambali, K., Shaharuddin, W. Y. W., et al. (2017). Human–wildlife conflict: A study of local perceptions in Jeli, Kelantan, Malaysia. *Malayan Nature Journal, 69*(2), 113–125.

Jamhuri, J., Edinoor, M. A., Kamarudin, N., et al. (2020). Higher mortality rates for large- and medium-sized mammals on plantation roads compared to highways in Peninsular Malaysia. *Ecology and Evolution, 10*(21), 12049–12058.

Kaffashi, S., Yacob, M. R., Clark, M. S., et al. (2015). Exploring visitors' willingness to pay to generate revenues for managing the National Elephant Conservation Center in Malaysia. *Forest Policy and Economics, 56*, 9–19.

Kingston, T. (2010). Research priorities for bat conservation in Southeast Asia: A consensus approach. *Biodiversity and Conservation, 19*(2), 471–484.

MNRE. (2016). *National policy on biological diversity 2016–2025*. Ministry of Natural Resources and Environment of Malaysia.

Mohd Rusli, Y., Radam, A., & Shuib, A. (2009). A contingent valuation study of marine parks ecotourism: The case of Pulau Payar and Pulau Redang in Malaysia. *Journal of Sustainable Development, 2*(2), 95–105.

Mohd Shahwahid, H. O., Mohd Iqbal, M. N., Amira Mas Ayu, A. M., et al. (2013). Assessing service quality of community-based ecotourism: A case study from Kampung Kuantan Firefly Park. *Journal of Tropical Forest Science, 25*(1), 22–33.

Myers, N., Mittermeier, R. A., Mittermeier, C. G., et al. (2000). Biodiversity hotspots for conservation priorities. *Nature, 403*, 853–858.

Ng, P. K. L., & Tan, H. H. (1997). Freshwater fishes of Southeast Asia: Potential for the aquarium fish trade and conservation issues. *Aquarium Sciences and Conservation, 1*, 79–90.

Nijman, V. (2010). An overview of international wildlife trade from Southeast Asia. *Biodiversity and Conservation, 19*, 1101–1114.

Noramly, G., & Yeap, C. A. (2001). Avifauna of Fraser's Hill. In A. Latiff, Z. Zuriati, M. I. Zaidi, et al. (Eds.), *Fraser's Hill: Physical, biological and socio-economic environments* (pp. 181–192). Universiti Kebangsaan Malaysia.

Nyhus, P. J., Osofsky, S. A., Ferraro, P., et al. (2005). Bearing the costs of human–wildlife conflict: The challenges of compensation schemes. In R. Woodroffe, S. Thirgood, & A. Rabinowitz (Eds.), *People and wildlife: Conflict or coexistence?* (pp. 107–121). Cambridge University Press.

Olson, D. M., & Dinerstein, E. (2002). The Global 200: Priority ecoregions for global conservation. *Annals of the Missouri Botanical Garden, 89*(2), 199–224.

Ooi, J. B. (1990). *Development problems of an open-access resource: The fisheries of Peninsular Malaysia*. Institute of Southeast Asian Studies.

Puan, C. L. (2013). *Integrating ecology into rodent pest management in oil palm agroecosystem*. Universiti Putra Malaysia Press.

Puan, C. L., Yeap, C. A., Lim, K. C., et al. (2014). Northbound migration count of raptors at Tanjung Tuan, Peninsular Malaysia: Magnitude, timing and flight behaviour. *Journal of Raptor Research, 48*(2), 162–172.

Puan, C. L., & Zakaria, M. (2006). Willingness-to-pay for the protection of a bird sanctuary: Perspectives from birdwatchers and non-birdwatchers. *Malayan Nature Journal, 59*(1), 81–92.

Richter, B. D., Postel, S., Revenga, C., et al. (2010). Lost in development's shadow: The downstream human consequences of dams. *Water Alternatives, 3*(2), 14–42.

Runckel, C. W. (2010). *A bird in the hand*. Retrieved March 26, 2016, from www.business-in-asia.com

Saaban, S., Othman, N. B., Yasak, M. N. B., et al. (2011). Current status of Asian elephants in Peninsular Malaysia. *Gajah, 35*, 67–75.

Saaban, S., Yazid, A. Z., Mustapa, A. R., et al. (2016). Human–wildlife conflict in Peninsular Malaysia—Current status and overview. In *Biodiversity forum 2016: Human–wildlife conflict mitigation and action in the agricultural sector*. Available via MPOC. Retrieved July 26, 2021, from http://mpoc.org.my/mpoc-perhilitan-biodiversity-forum-2016-human-wildlife-conflict-mitigation-and-action-in-the-agricultural-sector-22-24-may-2016/

Safari, S., Razali, N. A., Ibrahim, W. M., et al. (2021). From farm to China: A case study of Malaysian frozen whole durian export supply chain. *Economic and Technology Management Review, 16*, 1–20.

Siew, M. K., Mohd Rusli, Y., Radam, A., et al. (2015). Estimating willingness to pay for wetland conservation: A contingent valuation study of Paya Indah Wetland, Selangor, Malaysia. *Procedia Environmental Sciences, 30*, 268–272.

Sodhi, N. S., Koh, L. P., Brook, B. W., et al. (2004). Southeast Asian biodiversity: An impending disaster. *Trends in Ecology & Evolution, 19*, 654–660.

Syamsul Herman, M. A., Samdin, Z., & Shuib, A. (2013). Review of valuation from a non-market perspective: Travel cost method for rural tourism. *Worldwide Hospitality and Tourism Themes, 5*(4), 329–341.

Ten, D. C. Y., Jani, R., Hashim, N. H., et al. (2021). *Panthera tigris jacksoni* population crash and impending extinction due to environmental perturbation and human–wildlife conflict. *Animals, 11*(4), 1032.

Thorburn, C. C. (2015). The edible nest swiftlet industry in Southeast Asia: Capitalism meets commensalism. *Human Ecology, 43*, 179–184.

Tisen, O. B. (2016). Wildlife conservation vs conflict in Sarawak. In *Biodiversity forum 2016: Human–wildlife conflict mitigation and action in the agricultural sector*. Available via MPOC. Retrieved July 26, 2021, from http://mpoc.org.my/mpoc-perhilitan-biodiversity-forum-2016-human-wildlife-conflict-mitigation-and-action-in-the-agricultural-sector-22-24-may-2016/

Wong, E. P., Loke, V., Wadey, J., et al. (2018). Elephants, roads and drivers: Case study of Gerik-Jeli highway. *Jurutera, 9*, 26–29.

Wunder, S. (2005). *Payments for environmental services: Some nuts and bolts*. CIFOR Occasional Paper, no. 42. Centre for International Forestry Research.

Yeap, C. A., Sebastian, A. C., & Davison, G. W. H. (Eds.). (2007). *Directory of important bird areas in Malaysia: Key sites for conservation*. Malaysian Nature Society.

Zainol, N., Taher, T. M., Razak, S. N. A., et al. (2021). Wildlife crossings at Felda Aring-Tasik Kenyir Road, Malaysia. *Pertanika Journal of Tropical Agricultural Science, 44*(2), 401–427.

The manufacturer's authorised representative in the EU is Springer Nature Customer Service Centre GmbH, Europaplatz 3, 69115 Heidelberg, Germany. If you have any concerns regarding our products, please contact ProductSafety@springernature.com

Printed and bound by CPI Group (UK) Ltd, Croydon, CR0 4YY
01/10/2026
02226185-0003